Making Tobacco Bright

JOHNS HOPKINS STUDIES IN THE HISTORY OF TECHNOLOGY

Merritt Roe Smith, *Series Editor*

Making Tobacco Bright

Creating an American Commodity, 1617–1937

BARBARA HAHN

The Johns Hopkins University Press

Baltimore

The Johns Hopkins University Press
2715 North Charles Street
Baltimore, Maryland 21218-4363
www.press.jhu.edu

Library of Congress Cataloging-in-Publication Data

Hahn, Barbara, 1967–
Making tobacco bright : creating an American commodity, 1617–1937 /
Barbara Hahn.
p.   cm.
Includes bibliographical references and index.
ISBN-13: 978-1-4214-0286-4 (hardcover : acid-free paper)
ISBN-10: 1-4214-0286-6 (hardcover : acid-free paper)
1. Tobacco industry—United States—History. 2. Tobacco—United
States—History. 3. Tobacco industry—Virginia—History.
4. Tobacco—Virginia—History. 5. United States—Commerce—
History. 6. Virginia—Commerce—History. I. Title.
HD9135.H34 2011
338.1′737109730903—dc22   2011013289

A catalog record for this book is available from the British Library.

Special discounts are available for bulk purchases of this book.
For more information, please contact Special Sales at 410-516-6936 or
specialsales@press.jhu.edu.

Contents

Acknowledgments

Having had so much fun along the way, it would be hard to classify the process as work had I not needed so much help to produce the finished article. My doctoral committee nurtured my dissertation, paved my candidacy with good fellowship, and read my work with considerable attention and care. Peter Coclanis was the best possible doctoral advisor for one who has a tendency to bridle under the reins of control. I am grateful for his continued guidance and friendship, especially given how angrily I have sometimes fought social science history, which he so much prefers to stories. Few advisors could do so well. Alex Roland has been a terrific secondary advisor and a much-loved friend since we first met at Duke in 2000. He has changed my life several times over with his respect, friendship, and intellectual companionship. Alex also has always pushed me to find my own definition of technology. While I doubt he agrees with my definition, it remains true that we have discussed this project so long and through so many iterations that I cannot tell where my thoughts end and his begin. Nonetheless Alex, like Peter, always gave me my head. I very much appreciate the light touch of both men.

Paul Rhode also helped, from start to finish; I am grateful for his close attention, regular advice, and vigorous aid and argumentation. He took me playing in his favorite records, introduced me to the pleasures of the federal documents basement in the Walter Royal Davis Library of the University of North Carolina–Chapel Hill, and demanded as much rigor as I could muster. Long hours arguing about the nature of nature in the Carolina sunshine only clarified my thinking. He provided the agricultural census data from 1880 to 1940 that appear in my maps, and I am grateful for that—among so many other things. Paul has also read more of the manuscript more often than anyone. I doubt I have managed to persuade him. I hope his influence shows. Under his aegis, the Triangle Economic History Workshop always welcomed me, taught me to read economic history, and provided much camaraderie during the long years of lonely dissertating: thanks to all participants, Lee Craig especially. Lee provided the agricultural census data for the 1840–1880 maps, enthusiastically shared stories about tobacco cultivation, and patiently explained many complex economic theories. In addition, Gary Biglaiser supplied my readings in market microstructure, and Koleman Strumpf discussed both cigars and public choice. Many thanks.

Anonymous readers for the Johns Hopkins University Press helped a great deal. Hugh Gorman, Jenny Leigh Smith, and Lee Jared Vinsel each read the penultimate draft and gave useful advice. Paul Peterson cast a scientist's eye over my genetics interpretations. Bill Barney disliked this work in its dissertation form, from proposal to defense; I hope he likes it better now that I have found humans to carry my story of structure. I am grateful for his care and his honesty. Jacquelyn Dowd Hall always functioned as a fine cheerleader and impressive example. Thanks to both. Many other people in Chapel Hill, Durham, and Raleigh gave considerable aid: Harry Watson and Theda Perdue often served as my kitchen cabinet, my go-to guys on North Carolina and Southern history. Kathleen DuVal shared her notes at the Filson. Bruce Baker provided me with many newspaper clippings about Bright Tobacco in South Carolina. Carole Emberton, Malinda Maynor (and her father Waltz Maynor, who generously let me record his oral history), Paul Quigley, Rose Stremlau, M. Montgomery Wolf, and Tomoko Yagyu all made this a better book. So did Babette Faehmel. Adam Chill and Lisa Bartram (and Garp) provided me a gracious home while I worked at the Harvard Business School.

Joe Anderson, Mark Finlay, Shane Hamilton, Jenny Leigh Smith, and of course David Danbom and Claire Strom all helped make agricultural history a fun (and supportive) intellectual space within which to explore the themes that frame this research. Deborah Fitzgerald proved a tireless supporter of my work since I first met her at the Society for the History of Technology meeting in 2003; her efforts have done so much to animate agricultural historians and support and enliven our professional associations. Pete Daniel, Anne Effland, and Richard Follett aided the project in myriad ways. Rebecca J. Scott at the University of Michigan shared her enthusiasm for the peculiarities of tobacco, which rekindled my own at a crucial moment. Eliot Duhan, compliance guy, gave considerable advice on commodities and futures trading. In Berlin, at the Max Planck, Jonathan Harwood, Barbara Ann Kimmelman, and Tiago Saraiva made great comments. The SHOT community provided me with a wonderful intellectual home. Ross Bassett got me my first teaching gig at North Carolina State University. At the University of Cincinnati,

Wayne Durrill first sparked my interest in Southern history, and Mike Rhyne provided countless conversations on violence in postbellum Kentucky. Chris Phillips saved my bacon in the Black Patch Wars. Roger Daniels directed my M.A. and continued to provide excellent advice across my career. Zane Miller was a great friend.

I was so lucky to land at Texas Tech as my dissertation reached its conclusion. Start-up monies from the provost's office supported a summer back east, where further research helped me flesh out the arguments of the book, especially in its second half. The Junior Faculty Writing Group, organized by the extraordinary Lynne Fallwell, read several chunks and offered useful advice. My department chair, Randy McBee, and the rest of my colleagues made a supportive community out of disparate interests and agendas. Ron Rainger read a draft for my third-year review; his generosity and enthusiasm were remarkable. My students in Southern history, business and economic history, and the history of technology all made me read better and as widely as I had done in graduate school. Their questions and assumptions shaped my thinking on any number of points. I really have the perfect job, especially with the ability to teach so widely, in so many fields. I am grateful to the institution and to the people who compose it. In Lubbock, too, my medical team deserves thanks for making me face my family's genetic predisposition to cancer, thus far preventing my story from becoming an ironic addendum to this one. Dr. Anne C. Epstein, Dr. Catherine A. Ronaghan, and Damani Desai: thank you. Likewise, the Abouters keep me sane. I love you all.

Walter Friedman has nurtured this project since 2002–3, when the Harvard Business School funded my first explorations into the antebellum tobacco trade. His comments and advice over the years helped me choose the stories on which to hang my analysis, helped me decide where to publish what elements of the work, and helped me achieve what I sought from this overly ambitious project. His aid to business history more generally is well known to all who practice in the field, and his commitment to the discipline has helped me personally in my scholarly and professional life. Thank you, Walter— and Will Hausman, Roger Horowitz, Geoff Jones, Naomi Lamoreaux, Maggie Levenstein, Sharon Murphy, Larry Neal, Daniel Wadhwani, and so many other friends in the business history community. I could not have done it without you.

As a matter of fact, the business history community funded this project's research and writing for nearly a decade: the Harvard Business School, the late, great Newcomen Society, and the John E. Rovensky Fellowship all supported and funded this project over the years. So did the Business History Conference, the Center for the Study of the American South, the Filson Historical Society, the Virginia Historical Society, the history department and graduate school of the University of North Carolina–Chapel Hill, and the history department and provost of Texas Tech University. Individuals and institutions include the many librarians and archivists without whom this work would not have happened. Joe Schwartz at the National Archives and Records Administration in College Park let me wander the stacks. Others guided me through their collections: Laura Clark Brown and John White at the Southern Historical Collection; Frances Pollard at the Virginia Historical Society; and Janie Morris, Elizabeth B. Dunn, and Eleanor Mills at Duke. Entire staffs helped: at UNC–Chapel Hill, Perkins Library at Duke, Special Collections at NC State and East Carolina University, and Patricia Clark and Bruce Cammack at the Southwest Collection here in Lubbock. The entire staff of the Document Delivery Office here at Texas Tech and patient professionals at the Library of Virginia contributed mightily.

My family always cheered me on. Nancy Klein, Jerry Klein, and Melanie Roche gave more help than words can say. I am tearing up a bit as I write this. I am so grateful for everyone's help. Thank you all. So many people helped, but the flaws of this book are exactly and entirely my own.

Making Tobacco Bright

Introduction

The Marlboro Man evokes American identity. His cowboy masculinity calls up cultural memories of the Wild West. Riding a bucking bronco, wearing chaps over denim jeans, blue eyes nearly hidden beneath the brim of his Stetson, his rugged good looks express the frontier spirit. Individualistic, hard-working, self-sufficient, and manly, the Marlboro Man embodies virtues most Americans are proud to own. Yet there is another side to his image, another set of feelings it suggests. For many people, the Marlboro Man invites thoughts of cancer, the plague of modern society. Our fears of death and disease fuel anger at the tobacco industry. In so many settings, Big Tobacco works as the symbol of villainy. Movies like *The Insider* demonstrate the tobacco industry's manipulation of its customers into addiction. *Thank You for Smoking* extracts comedy from lobbyists who twist logic to deny the dangers of smoking. Tobacco industry executives testify in tandem that they never knew cigarettes could kill, even though the surgeon general first said so back in 1964. Slapped with huge settlements, Big Tobacco bears the brunt of acceptable public disapproval. If you use its products as recommended, they say, you will die. Few Americans today feel unalloyed pride in the Marlboro Man, Joe Camel, and the cancer sticks or coffin nails of that global behemoth, the U.S. tobacco industry.

Yet the Marlboro Man and Big Tobacco are relatively recent chapters in a very old history of tobacco in America. Cigarettes barely existed for U.S. consumers until the twentieth century, but tobacco reaches back to the very beginning. On first landing, Columbus received reports from his sailors that the natives smoked tobacco—perhaps to make themselves smell sweet? English settlers along the Chesapeake Bay and its tributary rivers made their first profits from the leaf and thus established the Virginia colony on a firm financial footing. Manufacturing tobacco emerged as a powerful domestic industry before the Civil War, a Southern industry that sold brand-name consumer goods to global markets. To the extent that they were known in that period, cigarettes were simply a paper-wrapped cigar—a novelty from Spain, a feminized cigar. Chewing tobacco, dipping snuff, smoking pipes, and cigars dominated nineteenth-century tobacco use. Cigarettes became the usual mode of consumption in the United States only around World War I. In those days, the American Tobacco Company employed branding and advertising strategies that had revolutionized the industry. At the turn of the twentieth century, cigarettes seemed modern, and the tobacco industry stood at the cutting edge of business practice.[1]

The cigarette century, as one scholar has called it, depended on a particular type of tobacco, the flue-cured variety called Bright Tobacco. Light and bright in color, mild in flavor, inhalable and suited to cigarettes, Bright Tobacco contrasted with colonial leaf that was mostly heavy and dark. Bright Tobacco first appeared in the southeastern United States, on the border between Virginia and North Carolina, on the inland Piedmont where the hills began their rolling rise toward the Appalachian Mountains. In the late nineteenth century, this kind of tobacco, produced with unique cultivation and curing methods, spread from its origin point in the Piedmont to new locations across the region and around the world, as consumption habits began to shift from chewing tobacco to smoking cigarettes, as the tobacco manufacturing industry became Big Tobacco. The rise of Big Business at the turn of the twentieth century saw tobacco manufacturers first competing and then combining into business alliances that carved up the world into markets for each firm, and cigarettes were the product they chose for the purpose. As a result, people in China, Brazil, and south-central Africa began to farm Bright Tobacco for sale to domestic markets and the global industry. Bright Tobacco can grow in so many disparate places because it has unique, recognizable, replicable characteristics that particular environments of cultivation and curing techniques will predictably generate.[2]

Despite these unique qualities, Bright Tobacco is genetically indistinct from other types of leaf. The varietal type and its peculiarities arise less from the seed than from the impact of cultivation methods on plant processes. Bright Tobacco is grown only in a very narrow environment: in light, loose

In the twentieth century, the American Tobacco Company used raw materials to advertise its cigarettes in the famous "Lucky Strike Means Fine Tobacco" advertisements. However, both cigarettes and the Bright Tobacco that went into them had been new at the turn of the century, a product of new tax laws, changing agricultural technologies, and the emergence of government science.
Source: *Life*, July 30, 1945.

soils that contribute to its final color, sandy and full of silicates, dirt that starves the leaf of the nutrients that could make it dark and bitter. It requires specific agricultural work, especially the techniques used to harvest and cure the crop. Unlike other kinds of tobacco, for example, Bright Tobacco is usually harvested one leaf a time. For burley, for example (a Kentucky type), farmers chop down the whole stalk to hang it for curing. Moreover, Bright Tobacco is flue-cured tobacco, by definition. It is cured by heat sent through ducts (flues) that protect the leaf from smoke and open fires. To some scholars, such technological distinctions seem called forth by the plant, as if "each variety demands a different cultivation routine." In truth, however, techniques are the products of human choices, whether the collective decisions gathered into the laws that limit what farmers do, or the individual decisions made during each crop cycle. The history of technology tries to explain those choices: Why do people harvest Bright Tobacco one leaf at a time? When and why did people take up flue-curing, and how did that make tobacco bright?[3]

Technology is a man-made thing, and so are its products. For that reason, this book is not quite a commodity study in the traditional sense—one in which an article's movements and impact, the products it becomes and the value added to it along the way, receive patient study along either economic or cultural lines. Instead, in this case, the commodity itself was under construction. Bright Tobacco lacked a clear identity for most of the story told in these pages. Historical actors barely recognized it as a unique, identifiable thing. How it became a category assumed natural, imagined to be a product of its seed, with specific traits revealed and procured by growing environment and techniques of cultivation, harvest, and curing—that is a story requiring centuries, and not one that aims inevitably toward its conclusion. Technological systems develop in concert with economic and social structures. The result is not clear from the start. This book therefore examines not the path of a trade good but instead the history of the technology that defines and produces that article—the causes of technology as much as its effects.[4]

The story of Bright Tobacco is but one example of the incredibly dense human histories that lie behind the world that seems natural but is instead technological. Human choices dictate the course of history. Weather and soil and the seeds are not responsible for the adoption of slavery, for the westward expansion of American settlement, or for the coming of the Civil War and the postbellum reshaping of the Southern economy. Nor can the sudden appearance of new machines account for the rise of Big Business at the turn of the twentieth century. Human decisions cause human history. Assigning causation to forces beyond human control limits our grasp of the past. It is a convenient way to distance ourselves from the consequences of our actions. This history of tobacco types demonstrates that sidestepping responsibility for history results in false stories, myths of discovery and invention. The history of tobacco technology teaches more than the social construction of tobacco types. It warns us to be cautious when our explanations turn to nature to explain human events.

THE REAL THING

To say that varietals are distinguished by technology rather than genetics likely rings warning bells for many readers. We live in an essentialist age, after all. Long-standing arguments about "nature versus nurture" in establishing human traits have tilted relentlessly toward nature as we learn more about genes from sequencing them, understand more about emotions and rational choices from advances in neuroscience and brain imaging technologies, and claim more about infant development from evolutionary theory. Homosexuals achieve civil rights by claiming that their sexual orientations are innate. A generation ago, common wisdom in the social sciences recognized that race was "socially constructed" and varied depending on the observer and the context. Nowadays, different medicines are shown to work differently on different races of people, even as race and the borderlines of racial distinctions remain poorly defined. The power these scientific efforts and approaches give us—to understand feelings and human relations, to treat diseases and track them among generations—has renewed contemporary faith in the explanatory power of nature. Yet, complexity dogs these efforts. Human gene sequencing has resulted in more uncertainty regarding disease causation. In tobacco genetics, further advances and elaborate efforts prove only more and more similarity among the established varietal types, even as seed breeders create ever-finer distinctions for economic purposes.[5]

Tobacco varieties are genetically very nearly identical. From the dawn of practical genetic analysis, geneticists have found the similarities among tobacco types shocking. In 1937, near the end of the time frame covered by this book, geneticists at the United States Department of Agriculture (USDA) discovered that the bright and dark tobaccos of Virginia were "so similar as to be almost indistinguishable." This lesson has been buttressed by more recent and sophisticated scientific study. Gene sequencing in 2007 demonstrated "the low level of genetic diversity within and among cultivated tobacco types," including not only bright and dark tobaccos of Virginia but also burley (long considered a separate line) and, to a lesser degree, oriental tobaccos. "Genetic diversity of tobacco germplasm was low, with a high level of genetic identity between the different types." The historical analysis contained in these pages, in other words, has come to much the same conclusion as the most recent genetic experiments. Readers seeking a closer analysis of the plant's genetics should turn first to the appendix. Others may rest comfortably in the knowledge that tobacco types are genetically so much the same that twenty-first-century advances in gene sequencing find further and deeper evidence of genetic similarity. The more precise the testing, the more pointed the results: the tobacco types are essentially the same.[6]

Tobacco types are legal distinctions, then, and what distinguishes them from one another is threefold: place of origin, technology of production, and market purposes. Where the leaf is from, what it is for, and what has been done to produce it: the USDA (founded 1862) uses these criteria to classify all the tobacco grown in the United States. Take, for example, the Connecticut Shade-Grown Cuban-Seed Cigar Leaf. Grown in Connecticut under shade (cloth tents covering the fields), entering the market only as the raw material for cigars, and from seed stock originally acquired from Cuba and bred for the state by the federal government, this type epitomizes today's tobacco taxonomy. The Eastern Dark Fire-Cured Tobaccos of Kentucky and Tennessee differ in significant respects from the Western Dark Fire-Cured Tobaccos of Kentucky and Tennessee, according to the USDA classifications. The subject of this work, the Bright Flue-Cured Cigarette Tobacco characteristic of North Carolina and Virginia, has also been defined along the same three axes. Between the Civil War and the New Deal, Bright Tobacco emerged to become the raw material of the cigarette revolution at the same time that the tobacco industry achieved near-monopoly.[7]

Bright Tobacco is therefore part of a classification scheme devised by the USDA. The USDA tobacco types fall into six major classes: cigar binder, filler, and wrapper; and air-, fire-, and flue-cured leaf. Note that some classes are defined by the purpose each serves (cigar binder, filler, or wrapper) and some by the curing methods employed to produce them. Flue-cured tobacco is a class that contains several types; these come each from a different area of the southeastern United States. To flue-cure tobacco means to control the temperature in the curing barn, to light fires at the mouths of the flues and thus expose the tobacco to heat while protecting it from smoke. The unique harvest method means that Bright Tobacco cures not on the stalk but as individual leaves. Leaves are harvested by priming—the method that plucks individual leaves from the plant as they ripen, one at a time, from the bottom up. Bright Tobacco has also been, for most of the twentieth century, sold usually at auction, in public warehouses, in small quantities: maybe a hundred pounds at a time, more or less. This distinctive cultivar is an artifact of this specific cultivation system, these technological processes whose result is a marketable product: Bright Tobacco.[8]

The morphological (visible, sensible) differences among the tobacco types need an explanation other than nature, seed, and genetics. The USDA classification scheme relies on region of origin, market purposes, and technologies of production, and the associations among these categories have a history. If one region grows leaf in a peculiar way for specific markets, that story can be chronicled. The emergence of tobacco varieties therefore relies on historical rather than natural phenomena. The distinct types evolved over time, in negotiations between buyers and sellers who both used the leaf's characteristics to determine its price: what is this leaf between us, and how do I know that, and what is it worth? Over centuries of trade, economic information about the goods fell into categories—the three categories listed above. These categories became the basis of regulation in the twentieth century. How different regions began to develop specific harvest, curing, and marketing technologies to produce commodities demanded for particular products—that process represents the subject of this work. How knowing where the leaf had grown meant knowing other important things about it—what had been done to produce it, its sensible characteristics, and what uses it served—is the mystery this work attempts to unravel.

APPROACHES

Why did the industry mechanize while agriculture did not? With this question in mind, initial research for this book focused between the Civil War and the New Deal, when the tobacco industry became Big Business and tobacco farms were small. Perhaps it was the harvest method: selecting and grasping individual leaves and tugging them off the stalk was a human action much harder to reproduce by machine than simply chopping down the whole plant and dealing with leaves later. So where did the priming method of harvesting come from? Some farmers claimed (in oral histories) to have adopted priming in the twentieth century, and New Deal bureaucrats found no standard method of harvesting Bright Tobacco. If priming was a new technique in the cigarette century, then cultivation methods seemed to have become more artisanal, more skilled and small-scale and regionally specific, exactly when the industrial side of tobacco production rose to dizzying heights of technological development and market control, evoking federal antitrust action in 1911.

Raw material production seemed to function outside the large-scale technological system devised by Big Tobacco for the manufacturing and distribution of consumer tobacco goods. Big Business did not, in this case, seize control of its inputs. Why?[9]

In the archives, too, the historical record of the period revealed an unexpected landscape: few sources talked about Bright Tobacco, or had any firm idea of tobacco types, or what techniques produced which characteristics in the leaf. For example, a merchant wrote to a planter in October 1879, "We hope you sun cured and flue cured your crop this year. You cant make any other sort of Tobacco that will buy you." There was no association, in this casual bit of correspondence, between curing method and the seed sown or the plant already cultivated. In a series of advertisements in the *Progressive Farmer* from December 1890, the Chamber of Commerce of Petersburg, Virginia, begged farmers who knew how to grow Bright Tobacco to come up there and help transfer the technology. This seems very late for that region's growers to still be learning how to produce the type. Then there was the letter from 1861 describing desired tobacco as "bright (not yellow)," while government experts even called parts of the burley crop "bright"—even though burley has long been counted as an entirely distinct variety, a different genetic "line" from bright. Bright Tobacco did not exist in the historical record in expected ways.[10]

It became necessary to construct a new story to make sense of these sources. If Bright Tobacco was not yet a kind in the late nineteenth century, with qualities assumed inherent in the seed, when and how did it become so? Evidence from earlier in the century supported the view that tobacco types were new at the turn of the twentieth century. Planter wrote merchant in 1846, "Tell what colour you want tobacco next year." Different periods had different practices. Did technology define the type? The question that research was intended to answer was shifting: how did tobacco become bright, and what processes made it so? Why did farmers use different techniques to produce tobacco in the colonial, antebellum, and postbellum periods? What did their specific methods make? Social constructivist approaches from the history of technology suggested how to look for the answer. Abandoning the notion that varietal types come from different seeds, that technology has developed merely to reveal the innate character of the plant, made considerably more sense of the historical record as it lay spread out in the archives.[11]

What made tobacco bright? What defined it as an article so unique that we ascribe its traits to the natural world, to the seed from which that commodity sprang? The answer to that question can still be found after the Civil War, when a series of institutional shifts in the economics of tobacco production and marketing brought all the tobacco types into being. Taxation, emancipation, and the USDA together made the tobacco types real. Tobacco types then became formalized in law in the 1930s, and the characteristics that define Bright Tobacco became solid enough that regulation could rely on their production. The story of tobacco before the Civil War, however, reveals exactly what was so new about postbellum production systems. The first few chapters, therefore, explain the different institutional frameworks of colonial and antebellum America and their impact on the tobacco trade. These periods demonstrate that tobacco types did not exist in this way—this association of seed with commodity, this confusion of cultural (and agricultural) techniques with nature—until after the Civil War, or even perhaps until the twentieth century.

THE STRUCTURE OF THE BOOK

In the colonial period, the inspection laws of Virginia shaped cultivation practices. Those techniques then provided the framework and set the limits of possibility on labor arrangements, farm size, and marketing methods. Delving deep into the history of this "Chesapeake system" of tobacco production, the internal workings of agricultural technologies, provides the basis for understanding the system's external, institutional, and economic causes and effects. By forbidding the sale of anything except first-growth leaves, inspection laws discouraged those cultivation and harvest methods that produced flowers, suckers, and ratoons—agricultural techniques that survived in other places, including Spanish Cuba and French Louisiana. In order to make the primary leaves as heavy and valuable as possible, the Chesapeake harvest occurred all at once, as late as possible. This created a peak labor demand that helped drive the adoption of slave labor in the region, to provide many hands at the critical moment. The cultivation system inspired by the inspection laws produced a type of tobacco recognizable in world markets, possessing characteristics guaranteed by colonial government inspectors and recognized in price differentials. The Chesapeake system of tobacco production proved durable even as markets widened with independence, although the freedom to meet more markets meant that new methods of curing, packaging, and marketing appeared as the eighteenth century became the nineteenth.

Virginia's colonial legislators had intended the inspection laws to boost commodity prices in an export economy, but the laws also laid the basis for a vigorous home industry as well, the subject of chapter 2. Colonial regulations established two different qualities of leaf, "shipping" and "manufacturing"— one intended for export, the other good enough only to stay home. This indicated two different markets, accustomed to different products. After independence, moreover, categories of leaf proliferated. Farmers attempting to get good prices meticulously prepared their tobacco for specific markets—by curing, preparing, and packaging it for a variety of buyers. Early manufacturers also used similar technological processes to create branded consumer products. Agricultural and manufacturing methods, from controlling moisture to preservation, resembled one another. Some production processes were modular: farmers, merchants, or manufacturers might stem or flavor the leaf, as the market demanded. As a result, the agricultural, commercial, and manufacturing sectors flowed into one another. Their boundaries were fluid. In this flexible, ambiguous industry, people continually refined and redefined the things they did to the leaf in order to claim more of its eventual value. Farmers and merchants might become manufacturers to preserve their stock, or manufacturers might become leaf dealers to dispose of theirs, as price signals dictated. People exploited the blurry boundaries between sectors in order to maximize their profits.

After the Civil War, three institutional shifts from the federal government transformed the tobacco trade and brought the tobacco types into being. Emancipation reorganized Southern agricultural labor from perpetual servitude to annual contracts, which steadily modified harvest, curing, and marketing techniques into those that facilitated the division of crops in December. Creditors needed technologies to change, and adopting fertilizer made it possible. Second, internal revenue laws distinguished manufacturing from agricultural and commercial processes by taxing manufactured tobacco. Revenuers found it difficult to know when a crop became a manufactured (and therefore taxable) good, so the laws they enforced eventually limited by licensing who bought and sold what form of the leaf. What farmers sold became, by definition, always an agricultural product. Finally, the USDA helped farmers meet their markets by establishing standard grades for their products. In doing so, the USDA borrowed a taxonomy from the 1880 census that had used place of origin, market purposes, and techniques of cultivation, harvest, curing, and marketing to distinguish types of tobacco from one another. In the twentieth century, these became the categories on which varietal designations rested. Varietal types then became the basis of New Deal production controls that told farmers not only how much tobacco they could grow but also what kind.

The Civil War therefore divides this analysis in two. When taxation helped drive the tobacco industry toward monopoly—or, more accurately, into oligopsony—postbellum farmers reacted by organizing politically. Chapter 4 examines the changing structure of agricultural production in the fields of Virginia's Southside and North Carolina's Piedmont regions, a consequence of emancipation, while chapter 5 takes up the agrarian movements organized along the lines of specific crops and commodities. The Alliance, the Populists, and their kindred associations have been studied more in terms of their unmet goals than their actual effects. Instead, this book describes the roots of twentieth-century identity politics in the farmers' agrarian radicalism. Farmers' advocates established a political base by identifying their economic interests with the agricultural sector—and, in the tobacco fields from which the Alliance sprang, they organized their constituencies according to the goods they grew. This book extends the history of farmer activism beyond its most common conclusion in the presidential politics of the 1890s, taking the story into the Black Patch Wars, a decade later, in the twentieth-century tobacco economy of Kentucky. Once farmers had accepted the types as regulatory descriptions of what they produced, their options in the marketplace shrank further. The result was violence, as the Black Patch exploded into burning warehouses, night riding, and murder.

Finally, the book turns to stabilization (a term borrowed from the New Deal agricultural programs): the series of laws in the twentieth century, culminating in the New Deal, that successively defined tobacco types and limited their production. While farmers and scholars alike believe that Bright Tobacco can grow only in very narrow conditions of soil and climate, the crop was spreading rapidly into new locations when, in the 1930s, New Deal production quotas designed to boost prices limited the crop to locations and farmers who already grew it. In that decade, the federal government established two contradictory sets of claims about tobacco varieties. In the first, the New Deal placed limits on farmers and what they could legally grow and sell based on what kind of tobacco they produced. The second claim was the recognition by USDA geneticists that the tobacco varieties were almost identical. The state and its institutions established an elaborate regulatory apparatus based on the structure of a tobacco taxonomy but then realized that this taxonomy had no natural basis. As a result, the peculiar harvest, curing, and marketing methods that had defined the Bright Tobacco type since the

New Deal then worked to prevent the mechanization of the harvest. The entire economic structure of land and labor had to change before any harvesting machine could work.

Every period of American history made its contribution to the creation of Bright Tobacco. For this reason, the social construction of tobacco types revises the narrative of national history. It makes the history of technology central to national history by exploring the causes of technological systems as much as their effects. It is a history of structure: what forces and individual choices establish those frameworks within which humans operate and make economic decisions? Technological change, too, stands between cause and effect, accelerating those forces that make it happen. Roads shape land values, but their routes are based partly on existing patterns of land use. From English settlement through the War for Independence and the War between the States, and on into the American century, tobacco played multiple roles. The themes of national history more generally associated with cotton—slavery and the westward expansion of plantation production, the diverging sectional economies of the North and the South, the coming of the Civil War and the transformations of postbellum Southern industrialization—appear as well in the history of tobacco. The development of tobacco into categories can help demonstrate how the world around us comes to seem natural, outside our power—when in fact it has a human history, caused by people making choices.

That being said, this book is not really about tobacco. Tobacco represents a single case study through which to examine the social construction of those historical forces that seem to lie beyond human control. Just because things are socially constructed does not mean that they are any less real. Class, race, and sex are prominent examples: social categories, peculiar to specific cultures and political economies, but inescapable nonetheless. The creation of a seemingly natural world can be illustrated through the social construction of tobacco types, an example of the human role in building frameworks in specific historical moments, structures that then appear as natural constraints, forces beyond human authority. Tobacco varieties exist, even if genetics can find little difference among them. They shape the choices farmers make, the brands Big Tobacco prepares, the deployment of labor in the fields and factories, the course of government legislation. The story of how they came to be is the story that lies before you.

Prologue

The slave was named Stephen. He was the plantation blacksmith, according to some accounts, or maybe the "trusted headman," at eighteen years of age. Despite this position of authority, Stephen was responsible for minding fires in the rain. But he fell asleep instead, according to the story. While he slept, the rain put out his fires—fires intended to cure a barn of tobacco. When Stephen woke, he stoked the fire high to cover his mistake. Some stories say he used charcoal and some say he used wood (access to charcoal can be explained by his blacksmith status, but wood better matched the curing methods common in the 1880s, when his story first appeared). When the barn was finally opened a few days later, it was found to contain "six hundred pounds of the brightest yellow tobacco ever seen." Variation in temperature during the curing process had produced an outcome planters had sought, according to some stories, since the seventeenth century: bright yellow tobacco. Thus was discovered the method of making tobacco bright.[1]

Technologies often have such foundational myths: Eli Whitney and the cotton gin offer one example, Thomas Edison and the lightbulb another. Add to the roster of great discoveries in American technology the story of Stephen, the sleeping slave, and his discovery of how to cure Bright Tobacco. Ignore the story's uncertainty about his fuel source. Ignore that he had no flues but rather stoked an open fire, and his fire likely smoked in the rain. Ignore all these elements of the technological system that would have resulted in a product quite different from that emerging from the Bright Tobacco barns of the 1880s. Accepting the myth requires ignoring many of its details. Some sources say it happened in 1839, some say 1840, and some say 1859. They say it happened on the plantation belonging to Abisha Slade, or maybe Elisha Slade—a Piedmont plantation near the border between North Carolina and Virginia. From that region, in the 1880s, Bright Tobacco was emerging—and that was when the story first appeared. We shall revisit it later, and its purposes will seem clearer when we place it in its context.[2]

Technological myths exist because actual histories of technology are extremely complex. Few scholars have the patience to explore the internal workings of methods and machines, although this is really the most effective way to elicit their causes and effects. It is difficult to grasp simultaneously the multiple, conflicting causes of technological change, its moment of closure, and likewise its various and contradictory effects. One scholar has described the role of such elements external to the technology as "the surrounding context that both animated and reflected that hardware," which captures the way technology stands between cause and effect. Experience has taught historians of technology the complexity and contingency behind supposed stories of invention and discovery. Yet it is vital that we replace myths of invention with truer stories of where technology comes from if we are ever to control it, rather than be controlled by its makers and adopters. Historians of technology have collected numerous case studies to demonstrate the intricacy of the procedures, to untie the knots and puzzle out the processes of technological change. Here is another such story.[3]

PART ONE

Tobacco before Types

[W]rite me if you have time something about Tobacco. The different kinds, what you think it will bring, and how it is selling in Richmd.

Isham [or Johann] K. Owen to William Thomas Sutherlin,
March 27, 1861

CHAPTER ONE

Making Tobacco Virginian

On June 9, 1785, out on Virginia's western frontier, Robert Wilson sent some tobacco to market on behalf of its growers. He sent it off to "the breaks," to the public inspection warehouses at Petersburgh, 130 hard miles away. There, government officials would break loose the staves of the thousand-pound hogshead and remove a few hands of tobacco from the compressed leaf within, a few samples to certify the tobacco's quality, to help find a buyer and set a price. The hogshead likely headed from there to London or one of the outports. Even after independence, tobacco usually still sold in the old imperial metropole, to be reexported to far-flung markets, or manufactured for sale to lords and common workers alike. First exported from the British Chesapeake in 1617, long-standing staple of mercantilist expansion, tobacco cultivation had spread settlers out along the river networks that fed the Chesapeake trade. Out on the river Dan, along the dividing line between Virginia and Carolina, where Robert Wilson got his neighbors' crops to the market, frontier farmers grew tobacco as the first English settlers of the tidewater had done: to establish their settlement, to profit the Crown, and to prosper.[1]

On this day, however, the merchant had to do more work than usual. One farmer, John Worsham, had 472 pounds to sell, but that was too small a quantity to market legally. He needed to fill a hogshead, which held at least a thousand pounds of tobacco leaf. English law excluded loose tobacco from the ports, and the Virginia inspectors assessed their fees by the hogshead. Transaction costs for hauling and freight and drayage, for turning up or reprizing, were usually counted at the same rate, by the hogshead. The giant barrel was an artifact of the institutional environment. Virginia's colonial legislature, the House of Burgesses, had spent a century experimenting with laws that tried to raise the price tobacco would bring. Since 1629 and the first downturn in price, the Burgesses had written successive waves of regulation. After 1730, they required government inspection to guarantee the quality of any leaf exported. Even after independence allowed direct sales to non-British buyers, the inspection system persisted. It was organized by the hogshead, however, and the merchant Robert Wilson therefore packaged Worsham's scant produce with that of another grower, a Captain Charles Williams. The merchant then split the costs of marketing the leaf, the inspection fees and transportation costs, the hauling and canal tolls, the prizing and the backhauling, proportionally between the two men.[2]

DEFINING TERMS AND IDENTIFYING PROCESSES

By 1785, the Burgesses' inspection laws had shaped a satisfactory technological system for producing tobacco in Virginia. This chapter examines this tobacco cultivation system in detail, looking for its causes as well as its effects. Studying the agricultural system will shed new light on some intractable questions in colonial Chesapeake history, from the origins of slavery to the impact of westward expansion. For some readers, the term "technology" may require the use of machines. In this book, technology will refer to a system of physical methods, tools, and techniques for getting work done. Where does a technology come from? Why do people do things one way, rather than any other? What options were available, and what inspired the choice of method? In the causes of technology, the organization of labor plays a role, as do the economics of production—the supply of inputs, costs of production, and demand for certain results. Economic decisions occur within frameworks of law and custom, however: a good example can be found in assumptions about what kind of work women can do, which of course affects the relative value of labor. This is one way that culture can shape economic fundamentals.[3]

The history of technology is rife with these kinds of contingencies. Contingency, however, eventually gives way to closure. Closure represents the achievement of a technological paradigm, an accepted method. It tends to look natural. The "Chesapeake system" for producing tobacco in Virginia achieved this closure in the eighteenth century. The tasks that produced tobacco leaves for market became so well established that they set the limits of growth—limitations on the size of farms, the scale of agricultural operations, and the labor required at certain points in the cultivation process. These limits were technical rather than natural, however. They resulted from human choices and institutions. For some scholars, making a hill for each individual plant represented the principal time constraint on cultivation. So what purpose did hilling serve, where did it come from, and when and why did it disappear? Likewise, harvest techniques and the preparation of the crop for marketing differed in the colonial, antebellum, and postbellum periods. The reasons for these differences, the causes of technology and its effects, are the subject of this work. This chapter examines the incentives and institutions that shaped Chesapeake settlers' technical choices. Other tobacco cultivation systems, which developed outside the region, will help bring both the causes and effects of technological paradigms into view.[4]

First, let us analyze the internal workings of the Chesapeake system of to-bacco cultivation, to provide the basis for understanding its external causes and effects. Growing tobacco included tasks that fell into four broad categories: preparation, cultivation, harvesting, and marketing. The processes varied from year to year, planter to planter, depending on individual conditions—a variable complex of tasks dependent on context and contingency, rather than a single, uniform practice. For example, clearing land of trees, stones, and wild growth created fields for the cultivation of any crop, not just tobacco. Not every farmer needed to prepare new fields every year. Tobacco cultivation tended to exhaust the soil, however, so clearing new ground was not a one-time process that built a farm out of wilderness and then ended. Farmers might sometimes grow tobacco on the same land some years in a row, or return to a field after several seasons lying fallow, or after rejuvenation by crop rotation, cow-penning, or other applications of farm-produced fertilizer. Similarly, cultivation techniques changed as farmers drew on experience and received letters from merchants who described the reasons each harvest brought its particular price. In their choice of practices farmers responded to such market signals, the land and its effects, the crop year's peculiarities, and the weather. Still, by 1785, farmers knew how to grow tobacco for market in the Chesapeake and grouped the tasks they performed accordingly.[5]

Preparation usually began with the seedbed. Selecting ground and clearing it, perhaps burning brush on it to kill weeds and their seeds, pests and their progeny, letting the ashes lend fertility to the soil, hoeing and raking the dirt—all these served to provide a baby bed to the tobacco seeds so fine that they needed mixing with heavier ashes to spread broadcast on the ground. Some cultivation instructions suggested covering the seedbed with brush to protect the seedlings from weather. Others recommended adjustable mats, cloth to pull back and let the plants take sun and strengthen. Then, the number and size of the leaves helped planters gauge the moment to transplant the seedlings into hills, mounds of earth about knee-high, flat on top.[6] Making hills in the fields as the plants grew in the seedbeds, through the winter and spring, was a technological constraint rather than a natural one. It involved correlating several tasks, so that the hills would be ready when the seedlings in their beds had grown to proper size and strength. Then they were ready for transplantation, as soon as the season permitted—which meant when it started to rain.[7]

TASK COMPLEXES

In tying hilling to the growth of the seedlings, planters engaged their workers in a task complex—a group of tasks that belonged together for technical, environmental, and economic reasons. Grouping tasks together was a technical decision that aimed farm work toward its ultimate goal, the production of commodities from plant processes. In this particular example, planters scheduled several jobs in tandem or in succession, so that seedbeds were ready when farmers saw in the rains of springtime the signal to start transplantation, and hills stood ready when seedlings had become hardy enough to move out into the fields. As a result, the agricultural calendar, the timing of particular tasks, became the product of decisions considerably more calculating than a simple observation of nature and its dictates. Decisions about one piece of work took the results of other tasks into consideration. Composite tasks such as seedbed cultivation and hilling appear again and again in agricultural technology. For example, seed and soil selection, seedling cultivation and hilling—and later, transplanting and pest control, weeding and re-hilling, cultivation and harvest, curing and marketing: each consisted of groups of tasks selected in light of tasks already done and the work planned to do next.[8]

Transplanting marked the transition from tasks of preparation to the work of cultivation. Setting out the plants into their hills lasted all spring, done only when it rained enough to really drench the earth. The job extended through the spring months because young plants often died on the hills, requiring "successive replantings, until the crop is rendered complete." The properly rainy weather for transplanting was called the "season." The term described the correct weather for several different moments in tobacco agriculture, stages when the level of moisture in the air became critical to the task at hand. As William Tatham put it, explaining to an English audience that had just lost its tobacco-growing colonies in a revolution, "when a good shower or season happens at this period of the year, and the field and plants are equally ready for the intended union, the planter hurries to the plant bed, disregarding the teeming element, which is doomed to wet his skin, from the view of a bountiful harvest, and having carefully drawn the largest sizeable plants, he proceeds to the next operation." With transplanting complete and every hill growing a plant, weeding and pest control continued together, probably performed at intervals that allowed work on other crops to proceed as well.[9]

Despite his portrayal of a planter rushing to the fields in the rain, doing work himself, Tatham described a transplanting procedure that required two people. One carried the basket full of plants and placed one on the top of each hill, while the other made a hole in the hill, planted the seedling, and built the soil up around the root "to sustain the plant against wind and weather." Although Tatham failed to note one common aspect of Virginian technology for staple crop production—its frequent use of slave labor—an instruction manual from South Carolina provided directions for stiffening the planter's resolve and providing for stern labor management. Recognizing that working outside in the rain all spring long might kindle some resistance, the instructions by Thomas Singleton warned, "the rainy weather must not induce you to call your negroes out of the field . . . embrace every rainy season, till your whole crop is set; then frequently [have them] go thro' your fields and wherever a plant is missing, supply the place with a living one." As these sources knew, many farmers actually managed a workforce rather than working on their own. Sons, family, neighbors, tenants, hired hands, or slaves: the Chesapeake system, once complete, involved making production decisions and getting others to execute them.[10]

One of the great benefits of this system to the planter was the way the organization of the fields allowed him to manage his workforce and calculate his production. By measuring the distance of the hills from one another, a set number of plants could be assigned to a specific number of hands—of workers: "care must be taken that they get . . . 18,086 hills as there are ten hands and the overseer at each place." Because cultivation had become, by the time of independence, a complex of predictable tasks including topping, priming, and suckering, these tasks made predictable the number of leaves each plant produced, so the work assigned and yield expected could be as closely correlated to the labor allotted as were the timing of hill building and transplanting. This spatial control of work and of laborers made production decisions more regular and aided the allocation of plantation resources. When the time came to weed and to worm, to top or to sucker, the number of hands necessary for how many days could be calculated by any enterprising planter. Cultivation tasks were controlled by the final outcome expected by the planter, and all aimed to produce commodity leaves from the plant and therefore directed the plant's energy to the moment of harvest.[11]

The task complex of cultivation included regular weeding. Workers chopped weeds down with a hoe, a particular kind of hoe, according to William Tatham, distinct from the hoe used in building the hills. Of course, this chopping tended to break down the hill in which the plant was growing, so re-hilling followed weeding, as night follows day. As one historian of agriculture succinctly stated, cultivating tobacco "consisted in alternately leveling and building up the hills." Pests dogged the plant at every stage of the process, from the black budworm that infested budding plants to the yellowish-greenish hornworm, four inches long at the smallest (but sometimes as big as a man's arm), plump and nasty, that appeared during the later work of weeding the fields. Killing the big worms by the thousands was, as Tatham put it, "a very nauseous occupation, and takes up much labour." Weeding, along with other variable cultivation chores, provides the twenty-first century scholar with the opportunity to grasp the decision-making quality of the task complex, both its causes and its effects. Weeding could be combined with worming, for example, as workers could "diligently search for the black worm which lives under the bottom leaves, and . . . must be destroyed."[12]

The mid-season chores of cultivation included priming and topping. These could be done at the same time or separately, or farmers could decide to do one or the other, or both, or neither. Priming meant removing those bottom-most leaves that planters called lugs. Lugs were bigger and thicker than the best leaves but often torn up and shredded during cultivation. Rain drove them against the ground, and weeding caused more damage. Yet, while lugs were less valuable than leaves, they did have some value by the nineteenth century. Priming lugs before harvesting the plant may have provided some needed funds at a cash-poor time of the production cycle.[13] Topping, on the other hand, pinched off the top of the stalk and prevented both the further upward growth of the plant and its reproduction. It therefore prescribed the exact number of leaves each plant produced and thereby contributed to the spatial control of labor—each hand responsible for a set quantity of ground, with a fixed number of hills, each with a plant bearing a specific number of leaves. By limiting the number of leaves, topping also maximized the size

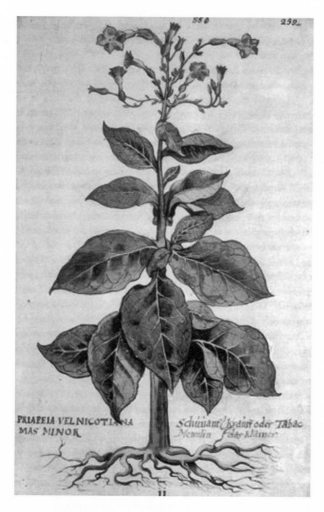

This botanical illustration of *Nicotiana tabacum* shows both its commodity leaves and some of its less valuable products—including both the flowering top of the plant and the second-growth suckers, which sprouted where the leaf grew from the stalk. The inspection laws of Virginia forbade the sale of both flowers and suckers and declared them unmarketable trash, although other European colonists and some native tribes treated them as choice parts of the plant. As a result, different cultivation methods developed in particular regions—techniques that maximized what different markets preferred.

Source: Leonhart Fuchs, *De Historia Stirpium Commentarii Insignes* (Basel: 1542).

of those leaves. Most planters probably let some chosen plants run to seed, however, in which case "topping" became a harvest of blossoms. Controlling plant reproduction helped the plant create what the planter sought: seeds for next season, or commodity leaves.[14]

Another part of the plant with little marketable value was called the sucker. Plucking these, or suckering, as some named the technique, followed topping as regularly as if the sequence were natural. These second-growth leaves appeared from the same juncture where the leaves sprang out from the stalk, and topping, the writers of cultivation manuals implied, made the suckers appear—as if inhibiting the plant's traditional mode of reproduction made it try to extend its life a different way. New suckers sometimes grew whenever one got pinched off. Some planters used suckers as a kind of fertilizer, but the second-growth leaves also might sell—but not in the eighteenth-century Chesapeake, as the inspection laws allowed the sale of only first-growth leaves. "This superfluity of vegetation, like that of the top, has been often the subject of legislative care," as had all means of producing second growths. Still, as Tatham remarked, "although this scion is of a sufficient quality for smoking, and might become preferred in the weaker kinds of snuff," its cultivation had been restricted by the Virginia legislators, the Burgesses, long before independence at the end of the eighteenth century.[15]

All these cultivation processes, in fact, aimed toward the work of harvesting, which began the curing and marketing tasks most directly regulated by the colonial legislature. Though weeding, worming, priming, topping, and suckering might continue through the midsummer season of cultivation, such drudgery served one ultimate purpose: to leave on the plants only the merchantable commodity, the first-growth leaves allowed for sale by the inspection laws of the eighteenth century. The goal of the inspection laws had always been to boost the price of tobacco, and, after a century of experimenting with controlling the quantity produced, the Burgesses in the eighteenth century took the easier path of controlling quality, which they did by forbidding the sale of anything other than the first-growth leaves. Therefore, the size of the leaves, their best and ripest qualities, and the threat of oncoming changes in the weather (or perhaps the impending harvest of some other crop) dictated when the harvest would begin. The planter combined, timed, and organized a selection of tasks his laborers performed. Then leaves remained the only things on the plant with marketable value. The ripeness of the leaves therefore dictated when harvest began.

Ripeness, however, was apparently not an easy thing to judge. Despite living in Virginia for twenty years, and despite his detailed expertise on every-thing related to tobacco production, William Tatham in 1800 still declared that he "would not trust my own experience without consulting some able crop-master in the neighborhood." He was not ashamed of not knowing for sure. He thought such a consultation was "not an uncustomary precaution among those who plant it." Because suckering helped pace the leaves' mat-uration, the emergence and removal of a specific number of suckers could determine the plant's ripeness. So could its appearance: leaves that looked "curdled and of a yellowish green cast" with the "points" turning down-ward to "appear somewhat shriveled" were adequate signifiers of readiness. Other appearance-based and color-based signals were also employed to time the harvest, and some of these were contradictory: some observers thought leaves "of a brownish colour," rather than a yellowish green cast, were the signal. Another apparently common way to gauge the moment best for har-vesting was to pinch or double over a leaf at a particular spot. If it cracked or broke clean, it was ripe, and the plant was ready for the knife.[16]

The fear of frost was a principal determinant in setting the tasks of harvest and curing in motion. Since the first-growth leaves remained the only thing on the plant with value, harvesttime became a calculation less about the plant than about the weather. As one overseer wrote a planter in 1810, "The To-bacco looks tolerable well provide[d] it was 'amonth erlyer in the year. [B]ut I apprehend the Greater part is to be cought in frost unless the fall should be uncommonly favorable. [I]t is but just now begining to ripen." As both owner and overseer knew, the growing process was flexible and the timing of other tasks adjustable—enough to push the plant's commodity production right up against the fatal change in weather. Frost would destroy the commercial value of the plant, although the plant could still be growing its valuable leaves to their ripest point even late in the fall. Weather continued to push on the tasks after harvest and associated with curing: rain pressed Landon Carter's workers into "getting the tobacco housed that was on the Scaffolds."[17]

Ripeness described a moment when the plant was most valuable; thereaf-ter, its appealing qualities began to deteriorate. After the harvest, sellers of tobacco leaves had to avoid plant decay in two directions—becoming moldy, or drying up so as to crumble and lack the body necessary for marketing and for the next stages of processing. Thus, the work of the harvest set in motion another complex of tasks, this one devised to preserve the plant by controlling its moisture in close relation to changes in the season and in the human-built environment, to make the commodity stable enough for storage and transportation. Weather was critical to how and when the procedures happened. Harvest began so late under the Chesapeake system of leaf cultivation that all its processes had to move quickly, and not only because frost might kill the plant just on the verge of yielding its principal return. The group of tasks that followed immediately on harvesting, too—the curing, in particular—needed rapid deployment and regular attention from the planter to ensure the stability of the cut plant. On the other hand, some of the work awaited a proper season, although not quite as wet a season as the one in which transplanting took place.[18]

METHODS AND MOISTURE

Controlling moisture: much of the technology of making tobacco products rotated around this activity. The moisture of tobacco was known as its "order" or "case." Case may have more specifically referred to how the tobacco would survive packaging and shipping, because moisture content helped dictate tobacco's preservation and therefore its value. Drying it entirely made it impossible to handle, bring to market, examine for sale, or prepare for its ultimate consumption. Farmers paid by the pound had incentives to keep it wet, too—but that made it subject to mold and funk. Anyone who has cut flowers to decorate a table, or harvested garden vegetables, or brought meat home from the market will recognize the peril of rot and decay. In terms of sales, of course, "a great deal depends on the good order of even indifferent Tobo." The moisture content of tobacco also changed in relation to the air around it, a process known in the tobacco trade as "sweating." Preventing or starting a sweat, or bringing one to an end, represented an important activity. Even so, tobacco would sweat twice a year, every fall and spring (which made its weight change), as in the merchant's description of desired tobacco: "It is prefered just in soft order enough to go through the sweat sweet."[19]

All the tasks of harvesting, curing, and marketing had something to do with preservation. Harvesting, wilting, sweating, hanging, and, finally, curing: each took a turn at slowing the plant's processes to the point where the final duties of marketing (striking, stripping, stemming, tying, and prizing) could wait for more leisure. These penultimate groups of tasks not only stabilized the plant but also established the color, flavor, and aroma of the final product. And, at least by the nineteenth century, in determining the price the tobacco brought, "much depends upon the Colour, & Substance." Plants well grown but ill handled during the many important stages of harvest, curing,

After they lost their leaf-growing colonies in the American Revolution, British authors published numerous volumes of instructions for producing tobacco. This engraving, published in London in 1800, illustrated the techniques of harvest, curing, and marketing as they had solidified in the Chesapeake during the eighteenth century. At top, tobacco sunned on the scaffolds outside the curing barn, while two workers prized (compressed) leaf into hogsheads. *Below*, a seated woman tied leaves into hands in front of a barn filled with cured tobacco. Then a worker rolled a hogshead to storage and, at bottom, inspectors examined the product.

Source: William Tatham, *An Historical and Practical Essay on the Culture and Commerce of Tobacco* (1800). Reprinted with permission of the University of Miami.

and marketing lost market value. Choosing techniques and gauging the stages formed decisive elements in establishing the commodity price: "make use of all opportunities, let slip none, for this is the critical time, as your crop depends on getting it safe into the house in good order." As in priming, topping, and suckering, technical choices both affected and depended on those that followed. Each job helped shape how and when the next one needed doing. In harvesting, curing, and marketing, however, instructions varied considerably, as did practices on the ground.[20]

Harvesting, curing, and marketing consisted of methods that had evolved during the seventeenth century. In the earliest years of tobacco production in Virginia, harvesting the plant meant cutting it down, chopping clean through the stalk near the ground. Curing consisted of simply "throwing the tobacco in piles, covering it with marsh hay and allowing it to sweat. In 1619," however, "a Mr. Lambert introduced the method of curing tobacco by hanging on lines, instead of sun curing in piles." When hung up and separated out in that fashion, the plants could dry out and stop sweating. A 1671 account describes driving a peg into the stalk to hang it for drying. In the eighteenth century, however, the method of harvesting became a bit more elaborate and replaced the peg with a harvest method that prepared the plant itself to hang for curing: "fixing your knife on the top of the stalk as it stands, force it down to the ground, one half the plant falls one way and the other half the other way, draw out your knife from between and cut each part off at one stroke." This left a few inches of stalk whole at the bottom, and the split stalk could then hang the plant upended over a stick.[21]

Harvest methods had evolved during the colonial period in order to reflect changes in the curing and marketing methods that followed. Even after independence, however, growers still had technical choices to make at curing time. Wilting and sweating, either before or after hanging for curing, remained possible options. Wilting apparently got the plant "limber enough to handle without breaking" and probably took place after the harvest, but before hanging. Thomas Singleton recommended that South Carolinians adopt the practice of turning the cut plant upside down on its hill to wilt it, then gathering it and stacking it on logs for sweating. This involved turning it "over and over until so limber that you can throw it about without breaking the stems." Those who wilted and scaffolded the plant this way should "get it off the ground as quick as possible, for without the greatest care the sun will scorch it while on the ground, which would be its total destruction" as "the heat of the sun and the moisture of the earth keeps the leaves in a sweat."

Tatham concurred that splitting the stalk when harvesting the plant allowed "a more free and equal circulation of air through the parts" that prevented "such partial retention of moisture as might have a tendency to ferment, and damage the staple."[22]

Singleton recommended this operation in order to make the tobacco "sooner fit for the house," by which he meant the curing barn. If wilted according to his instructions, the plant "cures much sooner, and the sooner and quicker your tobacco cures, so much the better for it." Yet some growers continued to employ sweating as one phase of curing, leaving the tobacco in bulk for some chosen time, as one author described having the plant "cut down and hung up to dry, after having sweated in heaps one night." Curing methods varied depending on the planters' needs and resources. So did the curing barns in which tobacco hung to dry. Tatham described four different curing barns, some with open spaces between the logs of the exterior walls, some sealed more tightly. He also explained that he had seen many other models. As for the interior arrangements of the barns, scaffolding from which to hang tobacco sticks outfitted them all. The plants hung from those sticks by the split stalk or, if necessary, by the join of the lowest leaf with the stem. Wilting and sweating both utilized similar devices, "scaffolds at your barn doors" for keeping the plants off the ground, closer together or farther apart, as preferred.[23]

Tatham explained in 1800 that "air is the principal agent in curing," but the historian Joseph Robert has found evidence for three separate means of curing by the nineteenth century: by air, fire, or heat, the last of which meant sending heat through ducts or flues to protect the leaves from the fire and its smoke. Choosing a curing method depended at least partly on the weather. The "crop-master" had to be "weather-wise," and "too much moisture" could be "tempered by the help of smoke . . . of small smothered fires made of old bark, and of rotten wood," designed to generate much smoke but little flame. If it were rainy, several sets of instructions advised the farmer to "put fires of green wood under it," but in warmer climates farmers would "seldom or never have occasion of firing, and the less firing the better, for it is hurtful to the smell of tobacco," as well as to its selling price. Tobacco well sunned on scaffolds before hanging in the house for curing might avoid fire in the curing process entirely. After all, Tatham thought, "[i]n this operation it is necessary that a careful hand should be always near: for the fires must not be permitted to blaze, and burn furiously," and tobacco hung poorly might fall and burn the whole barn down.[24]

Once the curing was complete, the leaf was all dried out. Planters and buyers could rely on the leaf not to rot. If handled in that state, however, the tobacco would crumble. It could not, therefore, undergo the last steps of preparation for market until it became more moist again, and for that, growers again had to await a season, "a rainy day" that was not only "most suitable" to the operation but also one in which "the hands cannot be so well employed out of doors." At that point the sticks could be taken down and withdrawn from the plants and the leaves stripped from the stalk. Sometimes marketing leaf involved stemming it—removing the thick central stem from the tobacco leaf, creating two strips. Tatham described stemming as a means of preservation, often employed when the weather put the tobacco "out of case." Maryland leaf appears to have come to market stemmed more often than Virginia leaf, but when Virginia settlers carried tobacco culture over the mountains into Kentucky by the nineteenth century, they often stemmed their harvest. This practice might serve buyers unwilling to pay as much for stems as for the strips made when stems were removed. Stemming indeed might classify as semi-manufacturing, a complex of tasks that will appear in a later chapter.[25]

Whether stemmed or not, individual leaves or strips were then tied into what Tatham called "little bundles" but generally went by the name of "hands," composed of several leaves, tied together at the stem with a bit of leaf or stem. The final task of marketing involved tying into hands and laying the hands in layers into the hogshead, the giant barrel for marketing tobacco that the inspection laws had established by assessing fees on that unit of sale. Layers were prized (an abbreviation for "compressed") into the hogshead for marketing. Prizing helped the leaf maintain its moisture level. If the leaf was too dry when prized into the hogshead, it might crumble during transport or upon removal for inspection or sale. On the other hand, farmers paid by the pound had incentives to sell their tobacco as wet as they could, but if "shipped off in a sweat, and it is well known a ship's hold will promote it: in that situation, the whole must perish . . . which the experienced inspector will prevent." Prizing required equipment, so a planter who owned the giant screw or a prizing beam might prize other people's tobacco into hogsheads for a price—or, a merchant could do it, as Robert Wilson demonstrated along the Dan River in 1785.[26]

Indeed, when the curing was done, the tasks related to marketing the leaf waited—for months or sometimes years. Once the leaf had been cured and all dried out, there was no rush to vend it. The pressure to complete tasks passed and selling waited on the weather, on the convenience of the planter

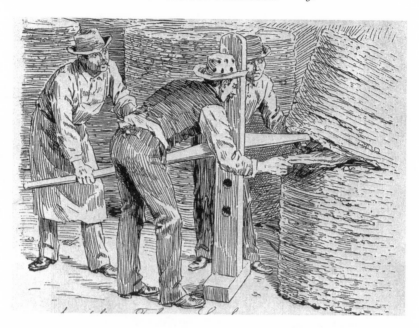

The legislators of colonial Virginia mandated the inspection of all tobacco intended for export, and other colonies (and later, states) established similar laws. This engraving, produced after the Civil War, illustrated the inspection process known as the "breaks." At the breaks, state government officials broke open the hogshead from around the pressed hands of tobacco and sampled them to guarantee their quality. As a result of the inspection laws, most freight and marketing fees were charged by the hogshead. In addition, since the inspection laws demanded that all tobacco sold be first-growth leaf, cultivation and harvest methods developed to produce that commodity from the plant. Harvesting first-growth leaves created a peak labor demand at harvest time that increased the incentives for plantation production and slave labor systems.
Source: Supplement to *Harper's Weekly*, Jan. 29, 1887.

---

facing other harvest tasks, and on buyers and merchants and market demand. The tobacco could rest in the barn while the planter turned to the harvest of other crops, the preparation of the land for the next production year, and the relaxation and celebration of winter holidays. Tobacco in fact seems to have reached the market sometime during the spring or summer after its harvest. Springtime brought the damp and rainy weather that suited the tasks of stripping and stemming—and besides, selling waited on buyers, whose ships arrived in the spring. The work associated with marketing, therefore, was a seasonal activity detached from agricultural processes. Robert Wilson, packing together two crops to send to Petersburgh in June, was brokering the crop of the year before. There is little evidence for an agricultural calendar that ended in December. Indeed, one scholar has calculated the colonial production cycle as a full eighteen months long from start to finish.[27]

As Robert Wilson demonstrated, sometimes a merchant rather than a planter did the work of prizing the leaf into the hogshead, and so the agricultural and mercantile sectors of tobacco production blurred together. In the earliest days of settlement, ship captains had functioned as merchants, sailing up the rivers seeking tobacco to trade for goods. This remained an option in frontier territories even after independence. Farmers could sign contracts to sell their tobacco "at the barn," too—a practice that will be explored shortly. The buyer rather than the grower, then, might take responsibility for some or all of the jobs associated with marketing. Stripping, sorting, stemming, and bringing into order—even tying leaf into hands and prizing hands into hogsheads—in that case became the work of the merchant rather than the farmer and his workers. In such a situation, too, the seller might avoid the inspection process altogether, although the buyer or middleman might then see the tobacco through the inspection process. Alternatives to export existed; the next chapter will examine them. Nonetheless, most growers—inheritors of the mercantilism that undergirded colonial settlement—used merchants as agents and exported their crops across the sea. For this reason, the inspection laws affected the technical choices they made.[28]

INSPECTION: ITS HISTORY, PRACTICES, AND EFFECTS

Inspection took place in public warehouses, often private barns that evolved under the inspection laws into public institutions. Public officials posted bond on their appointments to break open the hogsheads from around compressed tobacco, withdraw and examine samples of the leaf inside, and certify its quality. At "the breaks," as this inspection process was known, if the tobacco was of poor quality, trashy, sandy, full of stalks, broken, crumbled, or sweating and subject to mold on the voyage across the sea, or if it looked like the plant's second growths (such as suckers), the inspectors ordered it burned. Intended to raise the price of leaf by both limiting the quantity and certifying the quality of tobacco that reached market, these inspection processes represented the culmination of a century of legislative efforts. From 1621 forward, stint laws had tried to limit growers to a certain number of plants per "headd"; in the British colonies, however, assigning a quantity that each person could sow or sell ran into the question of who was a legal person—as women, children, servants, and slaves were not. Moreover, laws favoring folks who could command more labor, more servants, were hard to justify to a people eager for the possibilities the culture offered them. Virginia and its lawmakers went on experimenting.[29]

In the eighteenth century, the Burgesses took a different path to increasing the value of their export staple. Instead of directly controlling quantity, the new laws instead attempted to improve quality. They did this not by dictating agricultural practices but rather by forbidding the sale of second-growth tobacco of any kind. By outlawing the sale of second growths, the new law shaped the Chesapeake cultivation system into its familiar form. It also allowed more warehouses and granted negotiable receipts for the hogsheads that had passed inspection. These tobacco notes supplied a medium of exchange and attempted to make all tobacco of similar quality an interchangeable commodity. Colonial leaders claimed that the new system provided fewer advantages to big planters commanding lots of labor than their earlier efforts had done. Violent opposition eventually gave way, and government inspection of the colony's export commodity became common practice by about 1730. Whether or not the inspection laws succeeded at improving prices, they did create the incentives for what would become the Chesapeake system of tobacco agriculture. Proscribing the sale of second growths shaped the entire cultivation process—the topping and suckering and priming and curing as much as the sorting and tying and prizing that packed leaf into hogsheads for inspection at the public warehouses.[30]

The impact of the inspection laws on the Chesapeake system of tobacco production appears most clearly by comparison. Other methods of growing and harvesting tobacco, developed in other times and places, existed and persisted alongside the methods developed in Virginia. Comparing these cultivation systems to those of Virginia shows that none of the methods were inevitable or natural, learned over time as the one best way of extracting a commodity from the plant. Instead, different methods developed within the unique institutional frameworks of each region, tribe, and colony. Local selection among possible practices meant that different regions made different commodities when they grew tobacco by different means. Each system produced something unique to its locality, and each product served a specific market or met a particular set of demands. Supply and demand for particular qualities, created by specific processes and their unique growing environments, grew up together; comparing the characteristics of tobacco from different regions, the way each was made, and the purpose each served helps to illuminate the reasons for the Chesapeake system and its distinctive results: tobacco from Virginia, a first-growth leaf, harvested and cured on the stalk before being treated for marketing through the inspection system.

Native Americans in the Great Plains, for instance, had different institutions than the English settlers of Virginia and therefore developed different methods of producing tobacco. The Pawnees, a Central Plains tribe, did not cultivate the plant. They left the crop thick on the ground where the scattered seeds had sprouted and devoted their summers to food production instead. Rather than topping and suckering and then harvesting the leaves, they pulled up the whole plant after it had bloomed, then picked off the seed capsules and dried them separately from the leaves, considering the seed pods "particularly flavorful for smoking." The Mandan, Hidatsa, and Arikara tribes of the Northern Plains also preferred the reproductive parts of the plant to the leaves, and they especially liked to smoke the flowers. They grew individual plants in hills, as did the Chesapeake settlers, but their harvest began at midsummer, when the blossoms appeared. Their cultivation and harvest techniques produced several sets of blooms, harvested every four days. The species of tobacco employed by Plains tribes was *Nicotiana quadrivalvis*, and the Pawnees guarded their seeds jealously, while the Hidatsa tribe traded their products extensively with the Sioux. The ritualized use of tobacco represented a religious practice, and many tribes limited its cultivation, harvest, and consumption to male elders.[31]

Likewise, the French colonists of eighteenth-century Louisiana began their harvest at midsummer. Like the English who settled Virginia, however, they preferred leaves over flowers. Their harvest began when the suckers first appeared. They harvested the first leaves the moment that second growths appeared, allowing the suckers to grow to full size and then harvesting those when new ones appeared. To them, the suckers were simply another set of leaves—not the unmarketable trash that Chesapeake inspectors would burn. Harvest lasted all through the summer, then, in French Louisiana and permitted several sets of suckers to mature, allowing several harvests per year. Outside the inspection system, too, hogsheads were rare, and the marketing methods of French settlers reflected their long harvest of individual leaves. Some shipped leaf loose, or in boxes or casks, while others tied up leaves into shapes called "carottes." These options allowed the sale of small quantities, and the French government took all that its colonists made at prices fixed by the Compagnie or the *regie*, a government monopoly established to generate revenue from the colonies. "It was said that several cuttings were obtainable in a single year," according to Lewis Cecil Gray. "The quality of the product was reported excellent, better than that of Virginia."[32]

Well into the nineteenth century, even, alternative methods to the Chesapeake system persisted in other locations. Instructions for growing tobacco in Spanish colonies appeared to describe an entirely different plant than that of the Chesapeake, because they took advantage of different plant processes. The Spanish had been the first Europeans to adopt tobacco, as they had been first movers in so much imperial expansion into New World territory. Their methods, like those of the French, took advantage of the second growths that Virginia's Burgesses criminalized. The Spanish, however, cultivated tobacco in ratoons, another kind of second growth, familiar from the history of sugar cultivation. Ratoons appeared after the harvest and grew from the stump of the plant left in the ground. When Tatham had remarked that "this scion is of a sufficient quality for smoking, and might become preferred in the weaker kinds of snuff," he was referring to ratoons (known as "ground leaves" in England), although the refusal of all second growths in Virginia meant that he confused their several types, describing "an inferior plant from the sucker which projects from the root after the cutting of an early plant; and thus [produces] a *second* crop . . . from the same field by one and the same course of culture."[33]

Because Virginia Burgesses had forbidden the sale of second growths such as suckers and ratoons, the methods for producing them had disappeared in Virginia by the end of the colonial period. The technology for cultivating tobacco in the region, the Chesapeake system of leaf production, instead established the tasks here grouped into preparation, cultivation, harvest, and marketing—tasks such as seedbeds and hilling, topping and suckering, priming and chopping and scaffolding, and sweating and curing and prizing. Since only first-growth leaves would pass inspection and other tobacco would be burnt by government officials, the laws created an incentive structure for topping and suckering and growing those first leaves as big and heavy as possible and harvesting them all at once, at the end of the summer. The inspection laws that made Robert Wilson prize the crops of two growers together into one hogshead had shaped the entire agricultural system of production. In the way of technological systems, the different elements of production—the techniques, the labor and capital available, distribution to specific markets—all fit together into a coherent whole that looked natural, as if the colonists had learned the one best way to produce what imperial buyers sought. This seamless web makes technology look like a natural constraint, as when hilling limits farm size in so many interpretations.

Moreover, by requiring sale of only first-growth leaves, the inspection laws created a "peak labor demand" at harvest. The inspection laws therefore made slave labor central to the region's economy. Laws distinguishing slaves from other laborers had been on Virginia's books since the 1660s, and the availability of bound labor made possible the adoption of the abrupt, rapid, and labor-intensive complex of tasks related to harvest and curing. Once in closure, however, those tasks made larger workforces profitable and compelled their year-round employment. Thus, the harvest, curing, and marketing techniques stood between cause and effect. They took shape as a result of their external circumstances before making that context solid, as unchanging as a natural constraint. As Russell Menard noted in relating sugar cultivation to slave labor in seventeenth-century Barbados, technical change "sped up and intensified a process already under way." In assessing Chesapeake tobacco production, where 1730 marks the date of effective inspection laws, economic historians have identified a "steady upward drift" in the size of farms, and in the proportion of them using slaves, after 1720. What had been a crop suited to smallholders began to favor large-scale production. The reason for the Chesapeake planters' commitment to slave labor can be traced to the harvest technique, which itself can be ascribed to colonial market regulations.[34]

Thus, slave labor became an indelible part of the Chesapeake tobacco cultivation system at about the same time the colonial inspection system took hold. It would go too far to attribute Chesapeake slavery entirely to the inspection laws, but the timing presents evidence of more than coincidence: what had been a crop that rewarded smallholders became the staple of large-scale plantations nearly a century after its profitable adoption. The history of technology is often divided into internalist and externalist modes of analysis. The first consists of the examination of the techniques and tools, the organization of effort that composes the technological system of cultivation. External analysis examines the context for that system, both its causes and its effects. The inspection laws were causal. They provided a regulatory framework for cultivation, which created incentives for specific agricultural methods, which in turn created a peak labor demand at harvesttime. Large labor forces were crucial inputs to production under such constraints. The reasons for the method illuminate its effects. Close internalist scrutiny of the cultivation techniques clarifies the reasons behind their labor arrangements. Thus, technology stands between cause and effect, accelerating the very forces that account for its appearance. Understanding its impact and investigating its sources provides new answers to old historical questions, including the origins of Chesapeake slavery.

The inspection system also reorganized the marketing practices that had evolved over the years. At first, ships sailed the Chesapeake and its tributary rivers, filling their holds at plantation wharves before returning to England. Atlantic trade followed seasonal patterns to avoid the worm that ate wooden ships, so planters could somewhat predict sailing schedules. As settlement expanded, planters either consigned their goods for sale in Britain or sold directly at the wharf, depending partly on the difference between metropolitan and peripheral prices. In the eighteenth century, with the establishment of inspection, sales took place more at the inspection warehouses. Such business centers sometimes became towns—on Robert Wilson's frontier, Danville received both its name and its first warehouse in 1793. Danville also illustrates another eighteenth-century shift in marketing practices, as Scottish merchant firms established stores along the rivers of the Chesapeake, pushing settlement west from the tobacco coast. As Robert Wilson sent settlers' crops to Petersburgh for inspection before sale, the layers of middlemen between growers and buyers thickened on the American ground, while metropolitan firms in Britain diminished in number and increased the scale of their operations. By the time of independence, however, merchants in America had matured into a vigorous native community, and the first settlements along the Chesapeake Bay river system had grown into transshipment points for settlers farther west.[35]

Merchants mediated the markets in tobacco, providing credit for agricultural production and distributing its produce. Middlemen exist because they know more about markets than do the producers of goods, and more about the stock available than do buyers. A farmer selling a hogshead a year, or ten, or even a hundred hogsheads participated in fewer transactions than did the merchant handling the tobacco of many planters. Likewise, buyers relied on market mediators because merchants had more information about what leaf was available. This knowledge meant that merchants recognized the characteristics of tobacco, conscious of the supply of and demand for those qualities. Merchants realized more money for sellers because they knew who wanted what leaf. They found better leaf for buyers because they knew what existed. These middlemen needed their reputations to outlive any single transaction and therefore had an interest in fair dealing and the employment of accurate market advice. They matched buyers and sellers and provided information to

both, justifying prices by describing the goods. To sellers they described tobacco in terms of its purposes; to buyers, in terms of its origins. Sellers and
growers could accept a price when they knew what market niche their crop
would best fit. Buyers considered growing region a reasonable description of
tobacco's characteristics.[36]

Even the official government inspectors, however, could not tell one type
of tobacco from another. They filled out forms to certify the quality of tobacco they "passed," and these certificates made them choose whether a
hogshead was filled with Orinoco or Sweet-Scented, the "two great colonial
types." Although they opened and examined every hogshead of leaf rolled
down the waterways to the ports for export, saw every single hogshead legally exported from the colony, still the inspectors could not tell tobacco
types apart. As Tatham remarked, he had "not been able to learn from the
inspectors themselves (who I have frequently questioned thereupon) that
their botanical knowledge is sufficient to distinguish, at this day, one species
from another of the blended mass, by any leading characteristic upon which
they can pointedly rely." Growers and merchants attempted to claim certain
unique traits for their product, and sometimes they prepared and packaged
the leaf to call attention to these qualities—there is some evidence, for example, that Sweet-Scented more often reached the market stemmed than
did Orinoco—but these efforts were insufficient for official classification. Although "the *law* affects to make a distinction," the inspectors pretty much
just called it all Orinoco.[37]

WIDENING MARKETS DRAW DISTINCTIONS

All of this elaborate edifice, this culture predicated on the inspection laws, intended one result: higher prices from tobacco's buyers. In the colonial period,
of course, that meant the home country, Great Britain. Mercantilism assumed
that colonial production existed to benefit the imperial metropolis, the moth-
erland. For that reason, while the Burgesses wrote laws to boost leaf prices,
the empire across the Atlantic also regulated the market. The Navigation Acts,
for example, after 1651 outlawed sales to any but the British buyers and required colonial shipments in British ships. In exchange, the empire refrained
from growing tobacco anywhere other than the Chesapeake. This gave the
metropolitan British a brisk business in "re-exporting," in which Chesapeake
tobacco entered British ports before vending around the world. On the eve
of the revolution, between 85 and 90 percent of the leaf shipped to England

was re-exported elsewhere. France, particularly after the 1720s, bought Chesapeake tobacco from British merchants, taking about one-quarter of all the colonial leaf sold through England and Scotland. Holland and Flanders were likewise large customers for the leaf British colonists grew in the Chesapeake. Before independence, all these sales had to pass through the British ports en route to their purchasers.[38]

Independence meant the opportunity to meet those markets directly, but it did not prove easy. In the decades between 1820 and 1840, Britain still received 30 percent of all of Virginia's export leaf by weight, and 46 percent of its value (they got more of the good stuff). Further, re-exportation still accounted for nearly half the leaf that entered English waters. Britain also raised its duties on postcolonial American leaf from thirty pence per pound in 1783 to three shillings in 1815. Changing British tariffs across the nineteenth century shifted incentives for growers and buyers. British buyers paying increasing duties pushed growers to stem their leaf into strips, to avoid paying for the less valuable stems. When stems were needed, smugglers brought them in to England, creating a black market in the scrap used in certain products. As a result, stemming the leaf intended for British markets became profitable U.S. business, and the smuggling of stems proliferated. Such levies were only one example of British response to American independence. Attempts to grow leaf in new locations were another. Tatham's cultivation manual was an effort to transfer Virginia's tobacco technology to more hospitable holdings— Ireland, say, outfitted with a population fit to incorporate into the new version of empire.[39]

Virginia leaf continued to be popular in world markets, however, and some buyers sought direct access. Although France directly consumed only 8 percent of the U.S. exports, attempts to meet that market kept planters jumping. The seasonal appearance of French ships in the Chesapeake Bay meant reliable sales in the disordered decades after independence, but fashions in tobacco among the French kept changing. As one merchant described the situation, "The varities of the kinds they take are so great that a sample would be of very little use to you in sellecting." Finicky French buyers meant playing with curing methods in the effort to make "fancy colored tobacco," ranging from pie-bald or calico to green streak, or fawn, or straw, or hickory-leaf color. This was likely the reason one grower asked a merchant in 1846 to "tell what colour you want tobacco next year." In 1850, two agricultural surveys returned to the Patent Office revealed the spectrum of curing methods available, as a North Carolinian explained "that within the preceding five years curing had largely changed to a sun and air process," while a Virginia corre-spondent declared that "sun-curing had had its day and only a few were con-tinuing it." More variable even than in colonial days, nineteenth-century cur-ing techniques fluctuated often in efforts to create unique commodities.[40]

Even as other nations increased their direct purchases from the Chesa-peake merchants, European governments imposed regulations on tobacco to gain revenue from its importation. These institutions and taxes shaped the economic incentives of both growers and purchasers. Several nations estab-lished government monopolies on tobacco sales. France and Italy each had a *regie*, for example, that taxed imports and controlled sales to consumers. Austria, Spain, Sardinia, Parma, and even Mexico each extracted high duties from imports of American leaf. Nonetheless, political independence did widen the markets for the region's product. As Thomas Singleton wrote in the 1770s, instructing South Carolinians how to adopt the Chesapeake cash crop, "To-bacco will be in great demand for *France* and *Holland*: they have no resource but to us." New markets, however, required new preparations of the leaf, and widening demand after independence transformed the political economy that shaped leaf production. Independence changed market relations; the institu-tional framework of the trade encompassed new and different government regulations, which in turn affected the production and marketing systems of tobacco cultivation. The changing marketplace and the requirements of di-rect sales met innovative agricultural products and changing supply as the crop spread into new locations.[41]

CROP SPREAD AND TECHNOLOGY TRANSFER

Textbooks generally associate westward expansion with cotton production on nineteenth-century plantations, but tobacco prefigured Manifest Destiny. As Robert Wilson's settlement along the Dan River showed, representatives of Glasgow's tobacco lords helped colonists transplant tobacco culture ever deeper into the interior, following the waterways. Merchant-planters specu-lated in new land as well, spreading the crop to aid their investments, as set-tlers and squatters turned seeming wilderness into farmland. Colonization created capital by farm building. Rivers served as roadsteads that carried goods to and from the ports. Even during expansion, institutions shaped eco-nomic activity. The disputes over the Virginia-Carolina boundary illustrate the point: if the dividing line ran through the Nottoway River, it gave settlers river access to Albemarle Sound, whence tobacco could reach the sea unregulated

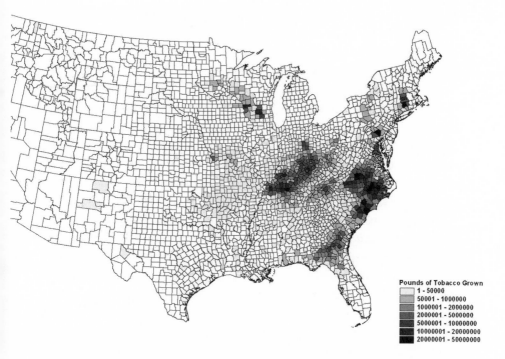

Agricultural production data gathered for the 1840 census, here compiled into maps, demonstrated the extent and location of tobacco growing in the United States at that time. As the map demonstrates, the Chesapeake Bay, Southside Virginia and Piedmont North Carolina, and the Kentucky–Tennessee border dominated the agricultural production of tobacco, but small quantities were grown in most counties.
Source: U.S. Census for 1840.

by Virginia inspection; a bit farther north and Carolinians would be rolling their hogsheads down the rivers that ran to the Chesapeake Bay. Virginia had long made efforts to prevent Carolina tobacco from entering the inspection system, because it might glut the market. After all, any tobacco that passed through Virginia's inspection warehouses became, in the eyes of the world, Virginia tobacco. It had passed the Virginia inspectors and thus possessed any characteristics the name implied.[42]

On the other hand, inhabitants of Virginia's northeastern neighbor, Maryland, responded differently to the Burgesses' market regulation. As in Virginia, Maryland's settlers clustered on the Chesapeake Bay, but its tobacco history diverged from that of Virginia. Early Maryland growers found it useful to occupy the market gaps left by Virginia's provisions. They packed larger hogsheads more densely than did Virginians, gaming the Burgesses' first efforts at limiting crop size. Apparently they also made a specialty of tobacco that differed from that of Virginia; some sources indicate they grew more Sweet-Scented as Virginia's product became more firmly Orinoco. Eventually, however, and at the urging of the Crown, Maryland in 1747 finally established inspection laws similar to those of Virginia. While grain production drew eighteenth-century Maryland planters away from leaf production, recurring high prices for the addictive cash crop meant that Marylanders occasionally returned to tobacco production. The reputation of the colony's products only rarely matched that of her Chesapeake neighbor, however. By the nineteenth century, Maryland tobacco had a unique identity in world markets, according to the price-circulars that listed tobaccos by their place of origin and clearly identified their differences by price.[43]

In the eighteenth century, too, Virginia's settlers pushed over the mountains, into Kentucky and the riverine west. They carried tobacco culture with them. At first they likely traded it amongst each other, but persistent settlement relied on links to distant markets and outlets to world trade. In the west, this meant the river system that emptied into the Gulf of Mexico. Spain, then France, claimed the Mississippi River at the turn of the nineteenth century, and European empires still hungered to profit from trade with the Americans upriver. After independence and even statehood, Kentucky tobacco still mostly floated downstream to New Orleans. The Mississippi River through New Orleans to the Gulf remained the principal conduit for Kentucky's leaf after the United States bought Louisiana. The city's merchants agitated successfully in the 1840s for a tobacco inspection system similar to that of Virginia; some dealers objected to specific provisions, but not to regulation itself, which aided quality. Indeed, those traders who desired a "particular regard . . . to the manner in which the Tobacco is cured and put up" declared that "what is raised by the Planters from Virginia . . . is by far the best." New Orleans's regulations on Kentucky leaf illustrated how durable the Chesapeake system of tobacco cultivation proved as the crop moved with settlement across the continent.[44]

Virginia's colonial inspection laws guaranteed certain characteristics in the leaf and had established the cultivation methods that created those traits. Political independence, however, contributed to differentiation among the leaves. First, new markets created variations in demand. Second, the spread of the crop to new locations modulated the supply. Supply and demand worked together within the emerging postcolonial institutions, as products grown in particular regions began to serve specific markets. The plant responds to its environment, after all, and new soils and varying cultivation methods contributed to distinctions in the leaf. Environment does not tell the whole story, however: in the mid-seventeenth century, some enterprising farmers grew in England itself a tobacco "of very fine quality" that sold in London as Spanish tobacco. In new locations, too, transferring technology led (as it usually does) to innovations in cultivation, curing, harvesting, and marketing methods. Firms trying to sell leaf in particular locations advised growers what those buyers wanted. A Kentucky merchant passed on the advice his Liverpool correspondent gave for "preparing tobacco for Africa," seeking "dark colour, with an appearance of richness, not prizy or Sticky, Supple in Condition to keep Sweet & Sound & free from mould." At the other end of the trade, African institutions provided incentives for leaf qualities and marketing methods.[45]

Independence destabilized production methods because world markets had eccentric preferences. Emerging markets demanded specific characteristics and preparations of the leaf. These modifications led merchants into new descriptions. Buyers wanted to know where it was from, and merchants supplied the information along with the leaf. North Carolina leaf that sold through Virginia before 1860 never had distinctive characteristics—or, at least, no price-circulars that listed it as a category have yet come to light. On the other hand, both Maryland tobacco and the "Western leaf" of Kentucky had distinct prices. Each therefore occupied a distinct category. Each had differences that buyers could recognize and cared to distinguish. To buyers in the world's markets, the growing region signaled specific leaf characteristics: a Copenhagen price-circular in 1805 distinguished between the "Tobac types" of "Marylands, Virgin, and Portorico."[46] Growers knew where they lived, however, and to them, merchants characterized leaf in terms of demand: some leaf might best "suit the African or West India Trade." Merchants likewise communicated cultivation advice to justify the price leaf brought: "The management of your crop in assorting, ordering & prizing was first rate. The quality would have been better had it been topped at 7@8 leaves . . . do this & you will gain more in weight than . . . be lost in [number]."[47]

Specific regions thus cultivated tobaccos in ways that suited particular markets. Individual importing nations had distinctive customs and regulations, and the emerging relationships between producing regions and market purposes inspired specific preparations and cultivation techniques. In addition, the inspection laws, established by the colonial legislators of seventeenth-century Virginia, had unintended consequences. By defining the sorts of tobacco acceptable to sell in world markets, first-growth leaves in particular, they created a category of tobacco that would not pass the inspectors but might be suitable for consumption at home. Indeed, evidence of price differentials between tobacco "passed" and "refused" at inspection abounds in the accounts merchants gave planters in the nineteenth century.[48] Moreover, distinctions between parcels for sale in Virginia often described them as "shipping" or "manufacturing" tobaccos: one hogshead might be "good dry Shipping Tobacco's & very sweet," while another from the same crop was "of a Manufacturing quality." Growers apparently knew the difference: "Common lugs . . . are very low, but shipping qualities and manfrs. are in good demand." For the Americans were not only producers of the leaf; they also consumed it—and how they transformed leaf tobacco into brand-name consumer goods for sale in both domestic and world markets is the subject of the next chapter.[49]

CHAPTER TWO

Growing the Business

On April 18, 1814, Charles Whittingham entered into an apprenticeship "on the art and business of a Tobacco Manufactory at Louisville Kentucky." Two years later he had grasped the "art and business" well enough to reassure his patron, a Boston merchant, of his investment in the tobacco manufacturer. "I think you have nothing to fear from him at Present . . . he has been very Fortinate in seling off his Tobacco, the funds of which he has Invested in Leaf Tobacco at a reduced price; and of an Excellent quality. He has on hand fifty H.hd.s of last years crop and 130 Kegs of Manufactured Tobacco, which have been put up since . . . December last." Moreover, Whittingham wrote, his boss had done especially well considering that "all the leaf Tobacco is principally Bought up. I fancy thare is not fifty hugheds for sale in the State of Kentucky. Manufactured Tobacco is seling as high as 22 to 25 cts by the Keg." The letters Whittingham sent back to Boston exhibit how much he had learned in two short years on the Kentucky frontier. He had acquired skills and knowledge in his apprenticeship, including an awareness of the commodity's availability, a manufacturer's concern with the outcomes of agricultural production.[1]

What did Whittingham mean by "Manufactured Tobacco"? Before the cigarette century, manufacturing tobacco was difficult to define. For our purposes, it will refer to those processes by which raw tobacco leaf from the farm became brand-name consumer goods sold in distant markets. Each part of this definition deserves some explanation. Branding consumer goods has generally been dated to the late nineteenth century; for this reason, early tobacconists led the way that other products—classic examples include soap and bread—would only later follow. Before the advertising era that ushered in the cigarette century, tobacco brands spoke mostly to merchants. Little evidence exists of U.S. consumer loyalty to anything like Marlboros or Camels. Users bought whatever plug or twist a store stocked. Indeed, even the term "consumer goods" in our definition may be misleading, as the final buyer might perform some of the work that readied the good for consumption. Most manufactured tobaccos looked like what today would be called "chew," but sometimes users crumbled it to fill their pipes. Different formulations likely served certain purposes better than others: a richly flavored plug probably contained too much moisture to easily smoke in a pipe, while many twists seemed made expressly for that purpose. Finally, to economic historians, sales outside the vicinity of production help distinguish manufacturing from home crafts.[2]

Given this definition, Charles Whittingham's factory may only have semimanufactured the tobacco it acquired. It may even have only stemmed the leaf, making strips for sale to manufacturers. After all, the Western leaf (as Kentucky tobacco was known) entered mostly export markets: Britain, Italy, or Africa. Although manufacturing for local barter was well known in the backcountry, part of the self-sufficiency of frontier communities, Kentucky tobacco had a "spongy property" and, according to an 1842 study, when "manufactured into 'lumps,' it loses its blackish rich color, and becomes, soon after exposure to the action of the atmosphere . . . what is termed 'frosted.'"[3] Since locally made lumps did not survive distribution, most Western leaf served export markets rather than domestic manufacturers. Western settlers floated their crops downstream, through the tributaries of the Mississippi River and down to New Orleans. European control of the Mississippi and its principal port did not prevent Kentucky growers from seeking the shipping lanes of global trade. On the other hand, Virginians developed considerable manufacturing enterprise in the nineteenth century. Virginia tobacco manufacturers sold brand-name consumer goods around the world. They made plugs and lumps and twists, packaged them for retail display and individual sale, and branded them for distribution at both the wholesale and retail level.

We know so little about this world. Scholarly treatments and popular perceptions alike usually trace the origins of the tobacco industry to the 1880s or 1890s. In that period, at the dawn of the cigarette century, the tobacco industry emerged in a new form, as part of the trend toward Big Business. The American Tobacco Company trust, the Dukes, and the Bonsack cigarette-rolling machine dominate the picture, obscuring the tobacco industry's true origins. Furthermore, historians, eager to explain the coming Civil War, have locked interpretations of the antebellum decades into a stylized national narrative, in which the North industrialized while the South committed to the plantation production of cotton using slave labor. The rise of cotton after 1820 interests most scholars more than the industrialization of the old tobacco trade. To be fair, some scholars have recognized the existence of the early tobacco industry, but few have examined its business history. They have deployed it as evidence for other arguments: to debate the compatibility of slavery with manufacturing, or the interactions between urbanization and the politics of a slave society. In textbooks, the antebellum South seems an agricultural foil to the industrializing Northeast, a nearly colonial appendage to the nation's dramatic economic growth, providing raw materials to the world's centers of industrialization.[4]

Scholars can be forgiven their myopia, however, when even their sources—including the authoritative Census of Manufactures—dismiss the early tobacco industry as "crude hand manufacture." The 1900 census described a tobacco industry that had originated in Connecticut in the 1850s. Census data themselves, however, tell a different story: of Virginia's manufacturing might, its dominance over a sector that made consumer goods and sold its brands to world markets. The antebellum census grouped finished tobacco products in different ways in every decade. Cigars were generally counted one-by-one, for example, while everything else was measured by weight; therefore, the best point of comparison between the two is their share of market value. By that measure, the 1900 census had the time and place of tobacco industrialization entirely wrong. The numbers tell the tale. In every decade between 1810 and 1860, Virginia tobacconists produced about *40 percent* of the entire industry's market value. Connecticut at the time of the Civil War produced less than *2 percent* of the industry's entire valuation. People all over the country made cigars. By 1860, however, Connecticut, New York, and other northern states made *only* cigars. Virginia led the industry for the half century before the Civil War, and the product of the Old Dominion was, overwhelmingly, "manufactured tobacco" (see table 2.1).[5]

Manufactured tobacco goods came in many forms. In addition to American and Spanish cigars and segars, there was snuff and pigtail, chewing and smoking tobacco, cut and spun tobacco, twists and plugs, lumps and rolls. The people who made these goods and the firms that distributed them referred to manufactured tobacco by brand name or maker, as well as by the object's shape, weight, and packaging. An 1831 product line bore the following description: "8s 12s & 16s Lumps in Boxes, and 13s Twists in Kegs Br[ande]d.

TABLE 2.1
The Antebellum Tobacco Industry

	1810			1840	1850	1860		
	Cigars $1,000	Mfd $1,000	% of U.S. total	% of U.S. total	% of U.S. total	Cigars $1,000	Mfd $1,000	% of U.S. total
California						76		0.3
Connecticut				2.1	2.0	562		1.8
Florida				0.2				
Illinois				0.2	0.5	148	62	0.7
Indiana				1.1	0.3	124	45	0.6
Kentucky				7.1	8.8	87	2,979	10.0
Louisiana	9		0.7	2.6	0.9	206		0.7
Maryland		200	15.8	4.0	4.5	697	72	2.5
Massachusetts		37	3.0	3.0	3.1	668		2.2
Michigan				0.1	0.7	66	6	0.2
Missouri				1.6	6.6	369	1,653	6.6
New Jersey				1.6	2.1	435	242	2.2
New York		45	3.6	14.3	11.0	2,404	1,718	13.3
North Carolina		0.2	0.0	3.3	2.3		1,117	3.6
Ohio				3.7	2.5	759	178	3.0
Pennsylvania	71	411	38.2	9.5	8.6	1,795	137	6.3
Tennessee				1.5	6.0	8	1,177	3.8
Virginia		469	37.2	41.4	38.2	109	12,237	40.0
West Virginia	—	—	—	—	—	—	—	—
Wisconsin						80	69	0.5
All other states		18	1.0	2.9	1.6	594	129	1.7
			99.5		99.8			99.8
Product totals	71	1,190				9,187	21,821	
U.S. total				1,260	5,820 13,491			30,802[a]

Source: U.S. Census of Manufactures for 1810, 1840, 1850, and 1860.

Note: The values of Cigars, Mfd, and U.S. totals are in uncorrected thousands of dollars. "Mfd" stands for Chewing, Smoking, and Manufactured Tobacco and Snuff. "%" represents a state's percentage of the total value of tobacco manufactured in the United States. Some of these states did not yet exist in the years covered, but they are included here to make this table correspond exactly with tables 3.1 and 5.1. In 1820, the varieties of manufactured products are too various to classify, and in 1840 and 1850, the census makes no distinctions between types of manufactured products.

[a] This total has been calculated and does not agree with the total presented in the census.

L. D. Jones."[6] The numbers referred to how many individual items made up a pound: fives were larger than eights, which were larger than twelves. Such details can illuminate both the business structure of the early tobacco industry and the technologies its entrepreneurs used. Ultimately, grasping the structures and production methods of tobacco manufacturing will indicate the impact of industrialization on agricultural production. After all, tobacco types proliferated in the nineteenth century not only to meet the export markets that widened after independence but to supply domestic industry. In order to understand the role industrialization played in the emergence of tobacco varieties, then, it makes sense to discern the origins and lineaments of the antebellum tobacco industry, which played such a crucial role in developing specific markets for distinct characteristics in the leaf.

TOBACCO INDUSTRIALIZATION: FIRMS
AND ENTREPRENEURS

We may never know who first manufactured tobacco in North America, or where the first factory appeared, or who first found a way to sell the goods outside the neighborhood. According to one authority, after Britain's Navigation Acts curtailed competition for Chesapeake goods, increases in freight costs led the colonists to manufacture their staple at home. Some sources describe tobacco factories on the Rappahannock River (a Chesapeake tributary) by 1732; at midcentury Landon Carter, Virginia tobacco planter, introduced amendments to the colony's inspection laws—changes concerning the export of trash, and allowing for the sale of seconds—laws specifically intended to protect domestic manufacturing. Henry Fitzhugh likewise described tobacco that "sells better in the country" and often sold hogsheads at home. In the nearly two centuries of colonial settlement, it is safe to say that Americans consumed tobacco as well as producing it, and they found ways to make and distribute goods for the purpose. In addition, Virginia's legislature changed the inspection laws in the first generation after independence, and the alteration provides another indication of domestic markets for the leaf. In 1805 Virginia inspectors no longer had to burn the leaf they refused to export. They recognized that this tobacco had a market at home, even if it could not be sent overseas.[7]

Still, individual firms and characteristic tobacco manufacturers provide structural examples of how industrialization began in the tobacco trade. There appear to have been several paths into tobacco manufacturing, each with its own supply chains. The rural path to manufacturing tobacco, for example, can be seen in the story of the brothers Peyton and Willis Gravely. Small-scale planters in the shadow of the nearby Hairstons (the largest family in Henry County, Virginia, and important members of the Chesapeake planter aristocracy), the Gravely family farmed, owned slaves, and kept a store.[8] They regularly bought small quantities of leaf from their neighbors, both black and white. Faced with a drop in the price of raw leaf tobacco, the Gravelys took up manufacturing in 1827, establishing a family business that would last into the 1880s and beyond. Like Robert Wilson, frontier merchant back in 1785, they dealt in quantities too small to sell through the inspection system. Perhaps

Many manufacturing methods developed in the blurry boundary between marketing agricultural products and industrial production of consumer goods. This 1750 engraving from the *Universal Magazine of Knowledge and Pleasure* illustrated early tobacco manufacturing operations that were indistinct from agricultural marketing processes. Clockwise, from the lower left-hand corner, tobacco was being removed from a hogshead, sorted, and then hung up to bring it into order (the technical term for controlling its moisture content). Two distinct devices for twisting and spinning tobacco appeared under the roof from which leaves hung while, in the right foreground, a screw press made tobacco into lumps. *Source*: "A Dissertation on Tobacco," *Universal Magazine of Knowledge and Pleasure*, Nov. 1750. Thanks to Bruce Cooper Gill of the Harriton Association, Bryn Mawr, Pennsylvania.

they intended to prize and resell it, but they first turned to manufacturing to add some value to the leaf, or to preserve it for later sale. They relied on local produce and made brands well known across the nineteenth century.[9]

This was apparently the same route into manufacturing followed by Hardin W. Reynolds. According to family legend and local lore, sometime around 1828 young Reynolds rolled a hogshead from home in Patrick County, Virginia, to Lynchburg, a ten-day trip. The low price garnered by all that work made him persuade his father to allow him to experiment with manufacturing twists, which he sold in South Carolina. According to a Patrick County history, "[n]early every planter who raises tobacco to any extent is a manufacturer"; some of these specialized more in manufacturing by buying unprized leaf at the barn, from their neighbors, as the Gravely brothers did. Hardin's nephew David Harbour Reynolds likewise followed the path that turned speculation in the raw leaf into manufacturing enterprise. He and a partner began business at Ward's Gap in Patrick County but moved to Patrick Court House about 1835. Their chewing tobacco sold throughout Virginia—at sites including Danville, Richmond, and Lynchburg—as well as down into Georgia. On the frontier they traded manufactured tobacco for country produce, groceries, and handcrafted articles, including woven cloth. Local historians remember the blurred boundary between agriculture and industry that took shape in the Virginia countryside as the young republic matured in the nineteenth century.[10]

The countryside nourished the manufacturing efforts of the Gravely and Reynolds families. Another route to industrial enterprise lay in the towns that sometimes developed around inspection warehouses, especially as they pushed tobacco cultivation inland along the waterways. Danville had received its name the same year as its first warehouse charter, in 1793—out in Southside Virginia, where Robert Wilson had packed together the crops of his neighbors. After independence, in towns like these, manufacturers might enter the business, part of the westward expansion of the crop. Danville and its agricultural hinterlands fed the industrial efforts of William T. Sutherlin, for example. Born in 1822, he worked his father's farm until the age of twenty-one, when he became a "leaf tobacco dealer" in town. A Whig of considerable means, Sutherlin became town alderman by 1851 and Danville's mayor in 1855. Sutherlin manufactured tobacco from 1849 or 1850, and for decades thereafter. He lived until 1892, and his business gave him considerable local power, as his presidency of the Danville Bank and his role in building the town's toll bridge attest. The man and the firm and the crop and the town grew up together, illustrating why urbanization so often functions as a proxy for industrialization.[11]

Cities could nurture generations in the tobacco trade, families that climbed an agricultural-commercial-industrial ladder into self-sustaining manufacturing enterprise. John F. Allen—the son of John F. Allen, "an old cigarmaker" of Richmond who had "always done a small obscure trade"—began manufacturing tobacco somewhat sideways. He began his career working for wages, an overseer in William Barret's Richmond factory. Typical of antebellum entrepreneurs, Allen entered into a number of different partnerships that covered many links in the commodity chain. He eventually turned the corner to formal and profitable manufacturing when he increased his access to the raw material. In early 1858, he joined forces with "a y[ou]ng man of not much m[ea]ns" who nonetheless, owing to his job with a German merchant firm buying Virginia tobacco, knew the leaf in the markets and the state of the trade. With access to such valuable information, the men began to make money steadily. By 1861, Allen owned a home worth five or six thousand dollars, and a few slaves besides, bringing his net worth to about eight thousand dollars. His career path illuminates the urban route into tobacco manufacturing, in which the trade grew from the export markets, and the towns like Richmond that thrived around inspection warehouses.[12]

Peter McEnery (1806–1881) also cultivated a tobacco manufacturing business in the tobacco towns of Virginia. It was his firm that produced the L. D. Jones brand, the eights and twelves and sixteens lumps he packaged in boxes, the thirteens twists sold in kegs. Like John F. Allen, he had over the course of the nineteenth century many partners in different firms, doing a variety of businesses within tobacco's complex commodity chain. One of his firms, the Richmond partnership of Henry O. and Peter McEnery, produced brand-name consumer goods before 1831. A generation later, McEnery had joined with James M. McCulloch to produce what became the well-known "Mac & Mac" brand of manufactured tobacco. That was not the full extent of his role in the trade, however. In addition, McEnery (with James Blythe Read) manufactured tobacco for the Petersburg commission merchants Branch, Winfree, & Company. That relationship made the firm a subsidiary of Thomas Branch and Company, a house of commission merchants in the classic antebellum Southern pattern, dealing in anything that came to hand, from real estate to commodities to slaves. McEnery's voluminous records demonstrate the intricate connections across sectors characteristic of the antebellum tobacco business.[13]

Robert Leslie (ca. 1794–ca. 1879) was another urban manufacturer, and one who blazed the trail from merchant to industrialist. His movement from commerce to manufacturing reveals a great deal about the early industry, not only because it occurred so early in the nineteenth century, but also because it makes clear the close relationship between mercantile activity and industrialization in tobacco's commodity chain. Leslie was what an older generation of scholars would have called a factor: "entirely a middleman," according to one taxonomy, but, more accurately, an agent of an overseas merchant firm. Born in Scotland, an early émigré to the young republic, he bought and sold Chesapeake leaf on behalf of the Dunlop firms (and others too) in London in

Liverpool. Having a man on the ground helped such metropolitan purchasers select leaf and collect payments; none seemed bothered when their agent established a factory in Petersburg in 1818. Like many tobacco manufacturers, he left behind a vast historical record that lies in several repositories, from the Library of Virginia to Duke University's Special Collections. Likewise, his multiple businesses and enterprises spread out from the agricultural production of the leaf as a varied but interconnected web on the Virginia landscape.[14]

Indeed, the path of the agricultural product, after curing and marketing, could go in so many different directions that a commodity web seems more accurate than a commodity chain. Borrowed from sociology and world-systems theory, commodity chains have become a popular subject of study in recent years. Today, goods sold in Wal-Mart (and other twenty-first-century retail markets) originate from many locations. Inputs and assembly wages cost less in underdeveloped nations, and poor countries fabricate the goods sold in postindustrial nations. Looking at historical commodity chains from the perspective of their origins rather than their end points, however—from the perspective of raw materials rather than finished goods—presents a different picture. Nineteenth-century tobacco is a perfect example: from the farmer's barn it was perhaps prized in hogsheads before selling in either domestic or distant markets, or perhaps sold at the barn and delivered in heaps on wagons. It might be stemmed after it was sold, or before; it might enter the hands of manufacturers, or be prepared as an agricultural product for a variety of possible markets. A web better represents the tobacco trade than does a chain—a commodity web spun off from agricultural production as the leaf made its way toward consumers in any number of possible configurations.[15]

MANUFACTURING PROCESSES

The boundary between agricultural marketing methods and industrial manufacturing processes was flexible and modular. Tobacco manufacturing was more processing than fabrication. The work of turning raw leaf into a finished, brand-name consumer good fell (as did agricultural production) into a series of task complexes: preparation, manufacturing, and packaging. Just as farmers did, manufacturers could pick and choose from the tasks in each group as they sought to produce consistent goods out of raw materials in a changing environment—including the weather, the markets and their demands, and the institutions of political economy. Some of the work done by manufacturers was identical to the work done by farmers. Controlling moisture mattered equally to growers, merchants, and manufacturers. Factory production therefore involved sweating both raw leaf and finished lumps, as well as activities to end the sweat. In another example of the blurry boundary between agriculture and industry, a manufacturer might buy leaf at the barn and, after selecting what he needed from the crop, prize the rest into hogsheads for sale, just as farmers did with their crops. In that case manufacturer functioned as merchant, just as Robert Wilson had done, although rather than taking a commission from a sale made on behalf of the grower, he would be a middleman who bought the goods and then resold them.

The groups of tasks that prepared leaf tobacco for manufacturing often overlapped with agricultural marketing: sorting, stripping, stemming, sweating, and bringing it into order. Sorting tobacco according to its characteristics of size, substance, color, its ability to take moisture (known as "chaffiness"), and the way its buyers predicted it would work through manufacturing processes all helped manufacturers decide what products to create from the available leaf. Tobacconists often made their top qualities first and then used the scrap in lesser products. After sorting, the tobacco usually needed to be made into strips by removing the central stem from each leaf, if that had not been done already. Sweating and bringing the leaf into order took place several times during manufacturing, since moisture was as crucial at this stage as it had been in cultivation and marketing. Indeed, the work of sorting, stripping, and stemming took place on farms as often as in factories. Moreover, just as in the agricultural tasks of cultivation, harvesting, and marketing, the industrial tasks of manufacturing and packaging blurred into one another. Each involved shaping the tobacco into plugs and lumps and twists, the finished consumer goods. Packaging the final outturn into caddies, cases, and kegs, too, involved sweating and shaping, as well as branding.[16]

Because the finished, manufactured tobacco had characteristics similar to the agricultural commodity, especially in its ability to take on moisture or dry out according to the season and as a result of its processes of preparation, its weight was variable. The product could gain or lose weight, just as could its raw material. This was likely one reason that the census counted cigars differently from all the various, multiple, competing consumer goods here called "manufactured tobacco." Even a journal promoting Southern industry, *DeBow's Review*, could say of tobacco in 1847 only that "its statistics, until lately, have not been very accurate." If modern historians know very little about the origins of the tobacco industry, its difficulty of measurement is one reason

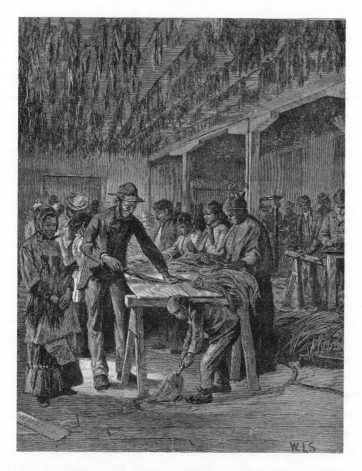

Control of moisture was a crucial part of manufacturing processes, just as it was in agricultural marketing—hanging leaf from the ceiling helped "bring it into order" as needed for processing. This image of a tobacco factory interior appeared after the Civil War, but before the cigarette revolution at the turn of the century transformed technological processes in use for a century or more in Virginia's tobacco factories.
Source: Supplement to *Harper's Weekly*, Jan. 29, 1887.

why. Industrial production overlapped with the processes and structures of agricultural marketing in other ways, too. Tobacco manufacturers made use of the business structures that served planters, especially the coastal commission merchants. Middlemen also served the tobacco industry, although they have been more closely associated with the plantation production of cotton considered typical of the nineteenth-century slaveholding South.[17]

Just as in marketing tobacco from the farms, manufacturing tobacco relied on controlling moisture at every stage, to prevent strips from crumbling in the handling and stop plugs from becoming moldy (or perhaps "frosted") in their packages. Leaf had to be dry to take flavor, which it absorbed in boiling vats of licorice or sugar or vanilla or tonka beans—the recipes used in "saucing" were secrets of individual brands closely guarded by their manufacturers, but the commonest ingredients were well known. Once flavored, the leaf usually had to dry out again to be twisted, rolled, pressed, or otherwise shaped before packaging. If saucing left the tobacco too moist, no effort made to dry it would keep it safe from mold through the time required for its distribution. One manufacturer instructed his overseer in language that indicated the extent to which manufacturing choices affected preservation: "Do not lay on the Sweetening too heavy as it is best they should pull up dry & stiff & be proof against mould particularly as any Cased Tobo. is very liable to mould in the South West where they may go."[18]

Even those processes specific to manufacturing—flavoring, shaping, and packaging the article for wholesalers or retail display—used machines similar to those of agricultural processing. The manufactured tobacco that dominated the industry's finished goods usually went through some version of the wing-screw press, a device that shared technological attributes with the prizing beam that compressed leaf into hogsheads. Both devices applied pressure to a flat surface that pressed the tobacco into the form in which it would sell, whether shaped by the hogshead of the farm or the tins that made individual lumps in the factory. By the 1850s, presses that used hydraulic and steam power had appeared—William Sutherlin, for example, adopted the innovation in Danville—but the hand labor of strong workers remained more common. The wing-screw press, whether operated by muscle or falling water or steam engines, could be outfitted with different molds, to make assorted sizes and shapes, distinctive products, as fashion dictated. One merchant described "Shapers" that he had found secondhand to make half-pound lumps that "run 24 lumps to the board with 8 boards to the set . . . they are made in the latest style and need nothing . . . to make them perfect for pressing dry work. With soft lumps they <u>might</u> spew a little, as all second hand and most new mills will do."[19]

Handmade rolls and twists likewise needed pressing into their packages, while the shape of lumps and plugs was more often formed by the act of pressing itself, between the shapers, the boards lined with plates. Whether pressing was part of manufacturing, or packaging, or both, the preservation of the

The wing-screw press was the most common device for pressing tobacco into lumps for
distribution to retailers and consumers. In this nineteenth-century broadside, the Eagle
Machine Works of Richmond, Virginia, advertised the range of presses they made and sold.
These tobacco factory fixtures were similar to the devices used by growers and merchants
to prize (or compress) raw tobacco into hogsheads for marketing (see p. 27).
Source: Kahl & Rahm, Eagle Machine Works, Broadside 18—. Used with the permission of
the Virginia Historical Society.

finished article and its presentation for distribution were both parts of the
manufacturers' work. In snuff, packaging helped define the finished product:
Maccoboy and rappee sold in earthenware crocks; dry Scotch snuff sold in animal bladders or esophagi called "weasands." Lumps and plugs usually wholesaled in oak boxes that could function as retail display for the storekeeper
who sold the goods to final consumers. Manufacturers were always looking
for oak planking of the right size and seasoning—"tobacco box stuff," as a
vendor referred to it—for packing up the goods. Wooden boxes took branding quite nicely. After packaging, the manufactured tobacco of course had to
be sweated again, to make it "perfectly summer proof and secure from any
Mould in any climate." The sweat house therefore appeared to be a common
feature of antebellum tobacco factories, sometimes kept separate from the
rest of production to prevent fires.[20]

There was little in the way of formal brand or trademark protection, so
some industrialists took out patents on their peculiar patterns and designs.
Most simply branded the packages as a way to make them seem unique, as Pe-

ter McEnery described in 1831. Brands could be proper names such as McEnery's L. D. Jones twists sold in kegs, or they might describe the product or provide fanciful associations. By 1860, a Cincinnati dealer had his "large and varied assortment" of stock printed on his letterhead, demonstrating that brand names outlived a single shipment. The brands he offered included Sallie Branch and Jno. M. Sutherlin, Crumpton & Paynes' Gold Leaf, Cherry Red, Luscious Luxury, and Wedding Cake, among dozens of other "Well Known & Popular Brands."[21] By midcentury, in fact, world markets recognized some of the American tobacco industry's brands as well as did this Cincinnati distributor. William Barret, for example, produced (in the Richmond factory overseen by John F. Allen) the tobacco known worldwide as Negro Head, produced from carefully chosen and richly flavored leaf. It appeared in English price-circulars, alongside another brand called Cavendish, and among the stocks of raw leaf.[22]

SUPPLY CHAINS AND DISTRIBUTION CHANNELS

Branding manufactured tobacco served the same purposes as branding other goods, but it happened earlier. Branding has generally been seen in the late nineteenth and twentieth centuries, when brands were applied to items otherwise indistinguishable: nails, bread, soap. Brand names convey information and claim differences among products. Amid the mass of lumps and plugs and twists, brands allowed buyers to distinguish a product from its competitors. Brands also helped manufacturers designate quality distinctions amid an array of products, to get higher prices for the better goods.[23] A brand guaranteed certain qualities such as flavor and moisture, but most American consumers had little choice: a store's ledger typically recorded only the sale of "4 plugs tobacco," not the brand, which the retailer chose. In fact, in this period, brands provided information to merchants more than to consumers. Distributors of consumer goods made the decisions about which goods to buy. At the consumer level, the manufactured tobacco seemed a commodity with few distinguishing characteristics: to most American users, it was twist, rather than a brand such as L. D. Jones or Negro Head. For merchants, wholesalers, and distributors, however, brands represented specific characteristics—flavor and aroma, size and shape, keeping capability, and the article's history in a particular market.[24]

Packaging distinguished leaf tobacco in the same way as tobacco brands, and raw materials often jostled with consumer goods on price lists. Virginia and Maryland tobacco came in hogsheads, of course, while Havana, Cuba, and Brazil tobacco sold by the bale. Florida tobacco appeared in cases while snuff sold in crocks or weasands, and Negro Head often appeared in tierces. Yet these recognizable categories sometimes contained even finer distinctions: York River was a category within Virginia tobacco, and at times it sold for a price between those agricultural commodities classed "part dark and sweet" and those described as "middling" or "fine and spinning" tobacco. Clearly it had qualities that distinguished it from standard grades of Virginia tobacco, qualities well understood by buyers, characteristics regularly produced by growers. Other nations and regions of the world also produced unique commodities in the leaf, and many also had industries that manufactured distinctive products. These also appeared in the merchants' lists and price-circulars, as Manilla Leaf and Manilla Cheroots brought different prices. In both the world and the domestic markets for tobacco, the line between agricultural and manufactured products was indistinct, as Kentucky strips might classify as semi-manufactured goods at home but as raw materials once they reached Great Britain.[25]

Manufacturing blurred into and out of other technological processes. The activities associated with manufacturing were modular; they could be done in bits and pieces. A small manufacturer might "do more in stemming," as did John F. Allen's father. At the same time, of course, a merchant who dealt in the agricultural commodity might conclude that "there is more Money to be made by putting up strips the coming year" than in selling raw materials.[26] Individuals slipped into and out of manufacturing as the profit margins shifted. Marketing raw tobacco did not differ significantly from manufacturing it: both controlled moisture, both compressed the leaf, and both tried to guarantee specific desirable qualities in the commodity. Manufacturers bought and sold tobacco readily, just as merchants did. Industrial production began with sorting, and some tobacco was discarded along the way: "I shall decline having more Tob manf/d for the present. . . . I should thank you to prize the Stems & the trash as refuse Tobo," said Peter McEnery to his leaf buyer, one hopeless year. When a factory shipped off "twenty-four Kegs manufactured and two Hogsheads leaf Tobacco" in a single transaction, it was unworthy of remark. The packaging helped define the tobacco contained within as either an agricultural or an industrial product, and manufacturers did business in both.[27]

Manufacturing tobacco was seasonal work, even more bound by weather and buyers' schedules than was agricultural production. As one merchant seeking stock wrote to William Sutherlin, "We do need fine pounds but prefer waiting till the season is favourable for manufacturing before we advise shipments." In another case, the manufacturer had to inform his New York merchants of established practice and tell them that despite their interest, "in the winter & spring it is almost impossible to put up Tobacco to advantage." Some of this seasonality had to do with weather, as moisture played such a crucial role at so many moments. Winter work proved "particularly susceptible to decay in the warmth of the following summer," according to one great scholar of the early tobacco industry. The weather in which tobacco was manufactured influenced its future quality as it made its trip through distribution channels: McEnery & McCulloch met a buyer's criticism with the explanation that "we used our best Leaf, & made a s[e]perate casing of it, & the dry cold Spring is alone the cause that it did not show to much better advantage . . . we defy any one to have done justice to Tobacco in such weather as has been experienced this Spring. [W]e will charge the failure to the weather."[28]

Some of the seasonality of manufacturing, however, must be ascribed to the emerging purchasing calendar, much as the season for selling agricultural products relied on the arrival of buyers. In antebellum commerce, retail storekeepers made spring and fall trips to New York City to stock their shelves. Credit reporters R. G. Dun & Company, founded in 1841, provided New York vendors with information about their customers' creditworthiness. The firm employed lawyers around the country to report what was known of each applicant's property and business, his personal habits and commercial connections, to determine what goods to sell him and on what terms.[29] Such wholesalers, whether on Manhattan or in the other major cities of trade—Baltimore and New Orleans, Boston and Philadelphia—made contracts with tobacco manufacturers to deliver fresh goods in certain seasons, in order to be sure of their supplies. Contracts to deliver the goods in spring meant "winter work," which began after the New Year, and manufacturers often scheduled their operations in order to work through sorted agricultural purchases in order of quality: "we are nearly through with our Spring contracts for No 1 work, will now prepare you some Ingraham lbs. & only wish that it had been in our power to attend to it earlier."[30]

The New Year marked a new season of work for another reason: the provision of labor, often on annual contracts. The tobacco industry was a Southern industry. Its workforce was largely enslaved. Tobacco manufacturers might have apprentices, of course, such as Charles Whittingham. He was learning the trade, however, not pressing lumps. Slaves did the physical work, hired by the year from planters who owned them. At the end of 1830, Peter McEnery postponed a trip for a few weeks, until "our Hands commence work," while a planter wrote William Sutherlin in Danville on December 7, 1860, about his promise to let "my negros Tim & Matilda" come home for the holidays, asking when the year's production would end.[31] The calendar of industrial production served well those manufacturers who were also planters: "little could be done in cold weather, and so when the factory shut down in winter, my father would take his large force of hands to the farm and put them to work." Manufacturers might own some workers and rent others, and some slaves possessed certain skills specific to the factories. Indeed, it seems as if particular tasks were the preserve of certain categories of laborers: urban factory hands received money for board and overwork, while the manufacturer paid "free Stemmers" by the piece or the day.[32]

Just as in plantation agriculture, labor management in tobacco manufacturing began to take modern, managerial forms in the decades after independence. Overseers were as useful in large-scale manufacturing as they were in substantial agricultural operations. Remember that John F. Allen started his career as an overseer in a Richmond factory; one of the enslaved workers in that operation was a man named Henry Brown, who escaped to the North (with the aid of abolitionists) by shipping himself to Boston by rail, in a box, as freight. Henry "Box" Brown's story, in the genre of nineteenth-century American literature known as the slave narrative, is one of the few first-person accounts of antebellum tobacco workers' lives. Raised on a farm, sent to the city after the death of his master, Brown described his working conditions in the Richmond factory under "the supervision of one of those low, miserable, cruel, barbarous, and sometimes religious beings, known under the name of overseers, with which the South abounds." The factory was a three-story space with nearly two hundred workers inside; although larger than the average firm, this scale of operations was not unknown to Richmond's tobacco industry, nor Petersburg's, nor those of other Virginia cities. North Carolina had some factories, too, although most were smaller in scale than those of antebellum Virginia.[33] Tobacco manufacturing was widespread in late antebellum Virginia and North Carolina—even more than tables 2.2a and 2.2b indicate, since so many farmers and merchants performed some manufacturing processes outside of these factories.

The use of slave labor represents another instance of blurry boundaries among the sectors. More types of business than plantation agriculture flourished in the political economy of the peculiar institution. Indeed, the scholars

TABLE 2.2
Antebellum Tobacco Factories in Virginia and North Carolina

	Virginia					
	Richmond firms	Richmond firms	Petersburg firms	Lynchburg firms	Danville firms	Danville firms
	1845 and 1850	1860 and either 1850 or 1845	1860	1850 or 1860	1850	1860
1–25 hands	4	1	0	16	17	22
26–50 hands	17	13	0	17	8	14
51–100 hands	8	29	11	3	4	3
101–50 hands	2	6	5	0	0	0
151–200 hands	3	0	2	0	0	0
Over 200 hands	0	0	2	0	0	0

	North Carolina									
	Caswell County		Granville County		Rockingham County		Stokes County		Surry County	
	1850	1860	1850	1860	1850	1860	1850	1860	1850	1860
1–24 hands	5	3	17	15	26	21	14	17	1	4
25–49 hands	4	7	0	1	1	4	0	0	0	1
50–54 hands	0	1	0	0	0	0	0	0	0	0

Sources: The data for North Carolina were collected from manuscript census returns by Joseph Clarke Robert, "The Tobacco Industry in Ante-Bellum North Carolina," *North Carolina Historical Review* 15 (Apr. 1938): 119–30, table on p. 125. The data for Virginia were collected by Suzanne Gehring Schnittman, "Slavery in Virginia's Urban Tobacco Industry, 1840–1860" (PhD diss., University of Rochester, 1987), table on p. 370. Because she was concerned with firm persistence, Schnittman gathered data only for firms that lasted over several years and traced them by name through manuscript census returns and city directories.

   Note: Data for North Carolina include only those counties having five or more factories, or factories employing more than forty-five hands. If all counties had been included, 20% of North Carolina counties had at least one tobacco factory. Burke County and Rowan County each had one factory in 1860 employing between forty and forty-four hands.

who wonder whether or not slave labor was consonant with manufacturing enterprise that used expensive machinery and relied on regular, willing workers should consider how "industrial" were the final stages of sugar production, and how agricultural were the first stages of tobacco manufacturing. The boundaries between primary, secondary, and tertiary sectors (to employ the language of economic analysis) have not always existed. The story of tobacco's commodity web allows examination of how such distinctions came into being. Farmers and storekeepers might easily slip into manufacturing, as the Gravely family did in Henry County; manufacturers bought leaf and sold scrap, sometimes for profit; almost everybody functioned occasionally as a merchant. Manufacturers added value to the good and broke it down into units for wholesale distribution and ultimate consumption, but the procedures they employed to do so were hardly unique to their sector. The methods a firm chose had little to do with formal distinctions between primary and secondary sectors, agricultural and industrial production, and more to do with the boundaries of and relationships among firms: what one bought and another sold.[34]

The entrepreneurial strategies of the men who manufactured tobacco reflected this reality. Peter McEnery's industrial works took shape partly as a subsidiary of the Thomas Branch commission merchant firms. McEnery not only manufactured tobacco but also bought and sold it, sometimes overseas, sometimes as an agent for others. Robert Leslie's close relationship with overseas markets provided him with an entry into manufacturing in the earliest decades of the nineteenth century; so did the partnership formed by John F. Allen, decades later. Many stories of the early tobacco industry, then, demonstrate the vital importance of merchant activity in American industrialization—and not just as middlemen. Although generally accounted as different sectors, the tobacco trade, with its complex web of connections, illustrates just how easily mercantile activity shaded into manufacturing, in exactly the way that Robert Wilson's actions in 1785, packing the crops of two neighbors together, performed industrial processes while simply mediating markets for growers. The fuzzy outlines of sectoral distinctions helped business grow. Infrastructural elements, financed by local leaders and their associations, proliferated in the antebellum tobacco towns: canals and railroads, warehouses and factories filled Shockoe Bottom in Richmond, for example. These allowed coastwise shippers to find Richmond a convenient port, as did the international tobacco trade.[35]

The histories of individual firms follow this model of blurred boundaries between firms and sectors. When William Sutherlin bought his neighbors' crops "at the barn," he sometimes used them in his factory, but he also often prized them into hogsheads and bundled them off for inspection and sale as a raw agricultural commodity. The manufacturer thus acted as a merchant, responsible for the product's quality and order when selling the goods. Yet even manufacturers stood as middlemen between growers and buyers, a link in the chain to consumers. Merchants were particularly well placed for early industrialization in the tobacco trade, as speculation in small crops might lead a storekeeper seeking profit into manufacturing. That was the story of the Gravelys, whose speculation in local crops led them into manufacturing by 1827. This historical sense of vague boundaries not yet formally divided into sectors receives support, too, from modern economic theory. Theories of market microstructure view supply and demand at the smallest level, as store managers lower prices in order to clear inventory. Technological processes were historically contingent. Even structures change. In the decades after independence, curing and marketing methods belonged sometimes to agricultural producers and sometimes to the industrial sectors.[36]

As in other examples of blurry boundaries between sectors, the same people and firms served both plantation agriculture and factory production, supplying raw materials and selling finished products. Merchants in the antebellum South, as in other times and places, mediated markets in more than one direction. They distributed the finished products of both farmers and manufacturers and provided inputs to both as well. A manufacturer wrote to his merchant describing the tobacco delivered by his firm as "sweetened & flavored much more highly than any Tobacco we have yet shipped," and in the next breath he discussed the supply of his inputs: "The last Tonqua sent . . . is of very poor quality . . . please send us some of the first really good Beans that you can yet hear of." Such a letter would be incomplete without some exchange of information regarding the agricultural commodity in which both correspondents were interested, although one manufactured it while the other wholesaled both agricultural commodities and finished consumer goods: "the quality of the crop we regret to say is thus far very indifferent," because, after all, the commodity had a sensible impact on the business of each: "useful qualities are selling very high."[37]

This tobacco manufacturer's missive, relic and reminder of the industrialization of the Old South, is as commonplace a piece of commercial correspondence as the archives hold. Its very typicality demonstrates how vast and untapped are the sources concerning early American industrialization. It reminds us that the tobacco industry was there in the 1850s just as it was in the 1830s and even the 1810s, nestled amid the commodity exports typical of plantation production, just as emblematic of the slaveholding South. Commodity production serving export markets had developed in the context of mercantilist European imperialism. It persisted through political independence and acquired new force with new crops, cotton especially. Just as merchants guided staples to their export markets, however, they also served as the middlemen between growers and tobacco manufacturers. As the agricultural exports of mercantilism required merchants, so did industrialization. "Credit supports agriculture, as the rope supports the hanged," says the French proverb, a reality recognized since the first century after Christ. As agriculture required credit, however, so did industrial production. Fixed sunk costs characterized both primary and secondary production, farming and factory work. Both required investments and were slow to pay returns. The differences between them are cosmetic rather than structural. Each takes time to make a saleable article from raw materials worth considerably less.[38]

MERCHANTS, MONEY, AND INFORMATION

Who financed all this? Where did the money come from for industrialization? What extraction or accumulation efforts capitalized those fixed costs sunk into factories and equipment, the rental or purchase of leaf and licorice and labor, boxes and brands? Such questions bedevil even the most-studied case of industrialization in the historical record, the British mechanization of cotton cloth production. Any exploration of the question should go far beyond the narrow definitions of inputs. Richmond became an industrial center not only because of its tobacco factories but also because of the infrastructure of tobacco inspection, the flour mills, the Richmond Dock and fourteen large-scale, privately financed storage warehouses that turned Shockoe Bottom into a coastal commercial center. Canals and turning basin and then the railroads connected the city to both its hinterlands of agricultural producers and stocks of available laborers, and also out to the wider world. On a day-to-day basis, however, the short-term credit that underlay production, that supported purchases and paid expenses as inputs moved through the manufacturing, packaging, and distribution processes, came from merchant networks—just as did the credit that supported commercial agriculture, the commodity production performed on plantations by workers who likewise constituted a fixed cost and investment by producers. Commerce paid for industry and agriculture alike.[39]

Merchants provided credit and distribution networks to manufacturers as well as to planters. They purveyed production advice and market information to both sectors, as a Baltimore correspondent wrote to Peyton Gravely in 1838: "You had better sweeten your fine Tobacco more than you have hitherto done." Since they distributed the finished goods of farmers, agricultural commodities by the hogshead, they also furnished the industry with its raw materials. Matching buyers and sellers represented one of their primary economic functions. As in the colonial period, they secured not only tobacco leaf but crucial information about the commodity. Without knowing what it was, how could one determine its value? The enlarging markets of the nineteenth century only meant more categories describing goods to justify their price. Independence from Britain meant a wider array of finicky markets not only overseas but also, as we can now see, at home. These widening markets contributed to the proliferation of tobacco types serving particular purposes. When a merchant described tobacco as "shipping" or "manufacturing," he was employing categories of quality and price that both buyers and sellers understood. It was the same with leaf classed as suitable for Africa or the West India Trade. Specific buyers desired specific qualities in the leaf, and merchants made a business of knowing what sold where.[40]

The informational work of merchants served to match farm produce with factories that processed crops into brand-name consumer goods. Finding the right tobacco meant identifying particular characteristics in the leaf. Over time, as farmers tried to meet market demand with the aid of merchant knowledge and advice, specific traits made in different regions and for specific purposes became well known and established. These distinctions were less about quality and more about qualities: one Virginian planter-manufacturer-storekeeper discovered on a trip to Georgia that "Luggs" were selling at "fair rates and there is good demand here now for common Tobacco but none for Fine."[41] When the price for fine tobacco dropped below the price for the more common product, something more complex than quality characterized the leaf. Categories of goods determined their price as much as did excellence. Knowing what an object is supposed to be helps one tell how good it is. Yet the categories that defined tobacco emerged only over time. Shipping versus manufacturing attributes helped growers to find markets and buyers to make choices among available stock. Then, widening markets of both supply and demand—new buyers, in both international trade and domestic industrialization, and new locations of production as the crop spread—continued the elaboration of important distinctions in the leaf.

Peter McEnery, for example, recognized in 1830 that a Baltimore merchant sought leaf that was "bright Yellow rich sweet & leafy." He traveled from Richmond to Petersburg specifically to pick up some hogsheads for his correspondent "the most suitable for your mkt." Bright Tobacco was becoming something of a mania, although how it was made mattered not, and even its characteristics were still under construction. In most cases it seems that the term "bright" referred to the color of the leaf, but sometimes even that much was doubtful. As late as 1861, a merchant requested that a manufacturer send "some good Com[mon] Pounds like the 'Fulton' with bright (not yellow) wrapper."[42] Demand for light-colored leaf represented a change. The quality of leaf in colonial Virginia came from its darkness, richness, and heaviness. Nannie Mae Tilley, the great historian of Bright Tobacco production, may have dated the "quest for yellow tobacco" from 1606, but most colonial producers (including Henry Fitzhugh, eighteenth-century Chesapeake planter) viewed tobacco as good when it was heavy and dark. After independence, with the development of the domestic industry, calling the leaf "working Tobacco" no longer sufficed to distinguish it. Color mattered in identifying the desirable characteristics of the agricultural commodity: leaf could possess "fair color though not bright enough for fine."[43]

The novelty of the desire for light or yellow tobacco can further be seen when we catch up with Charles Whittingham, still at work in Kentucky tobacco production twenty-five years after his apprenticeship began in Louisville in 1814. As Whittingham's indenture demonstrated at the start of this chapter, tobacco processing came to the state almost as early as tobacco cultivation. Louisville was in the central part of the state, however, part of the fabled bluegrass, at the falls of the Ohio River where commerce flourished. The movement of tobacco production into the more western parts of Kentucky provides an example of the proliferation of types (and the importance of merchant information) that followed the twin transformations of westward expansion and independence. Whittingham participated in these processes through the interrelated firms and sectors of tobacco's commodity web. William Barret, whose Richmond factory produced Negro Head under the oversight of John F. Allen, had a brother named Alexander. Alexander Buchanan Barret married the sister of David Bullock Harris, who owned a stemmery at Cloverport, west of Louisville. Harris and Barret traded even farther west, centering their activities around Henderson, and sometimes traveled even into Missouri to supply the factory. By the 1840s the firm exported Western Leaf directly to the British merchants Gilliat & Co., of Liverpool and London.[44]

In the years between 1838 and 1841, the men had worked at setting up their business, buying and hiring laborers, having their Louisville man—Charles Whittingham—ship to the Cloverport factory both hogsheads of leaf and "Bars Iron for Tobacco Presses with Screws &c." In the early years, these tobacconists described leaf in generalities, or in terms of its purposes: "good Stemming qualities," "Strong Tobos," or "Strong Leaf."[45] By 1843, however, Barret was remarking that color was "soon to be the sine qua non" in distinguishing quality; different colors of leaf would soon indicate qualities that served specific markets. "You should Stem the most leafy & bright Kinds & prize all other sorts," Barret buying in Henderson wrote Harris at the Cloverport factory. In another letter from the same period, Barret quoted at length from the Gilliat circular instructions for preparing leaf for Africa; it was he who described the leaf desired in Africa as of a "dark colour, with an appearance of richness, not prizy or Sticky, Supple in Condition to keep Sweet & Sound & free from mould." He intended that his careful copying of Gilliat's instructions should help his partner produce and package the tobacco just as the market desired.[46]

The merchants' letters expressed important lessons: light tobacco and dark tobacco, both from western Kentucky, each served different purposes. They entered different markets—dark leaf, prepared in specific ways for export to Africa; light bright leafy tobacco, stemmed for British buyers, or perhaps Virginia manufacturers. Putting leaf with different characteristics into categories based on who wanted it helped farmers and dealers better meet the demand. Such advice also began to delineate the characteristics of specific types serving particular purposes, even if those types came and went with shifting fashions. David Bullock Harris wrote John Gilliat & Co. in London about a popular Kentucky product, for example: "a description of tobacco much liked in Orleans . . . sold there under the denomination of 'Yellow Bank Tobacco.'" Grown mainly where Henderson and Daviess Counties met, Yellow Banks came from northwestern Kentucky, near the corners of Illinois, Indiana, and Missouri.[47] Its momentary prominence ensnared Peter McEnery, the Virginia manufacturer of several decades' experience, into a leaf-dealing episode that cost him dearly. In 1847, he began to notice that buyers in New Orleans displayed a decided preference for Yellow Banks tobacco, and that it was selling at a higher price in New Orleans than in London.[48] Merchant as well as manufacturer, classic player in the antebellum tobacco trade, McEnery took the plunge.

In May 1852, when the price of raw tobacco was low and the demand for manufactured tobacco high, McEnery placed an order with his New Orleans merchant-correspondent to buy 150 hogsheads of the Yellow Banks from Kentucky, and he specified purchases of a quality "suitable for fine cut in New York."[49] McEnery was buying Yellow Banks in New Orleans at six and a half cents a pound, while commanding in New York between seventeen and twenty cents per pound for the products he manufactured from it. By September, his price calculations—along with the likelihood of "very bad prospects for the Western Tobacco crop"—pushed McEnery and his partners to consider expanding their Petersburg factory to process the goods.[50] By November, however, the price difference between New Orleans and New York had disappeared. Worse, the fifty hogsheads of Yellow Banks that McEnery had shipped to his correspondent in New York appeared unlikely even to find a buyer. All was lost and the firm dangerously overextended. By December 20, as the firm completed its contracts to deliver manufactured tobacco and closed out its business for the year, McEnery and McCulloch discovered they had to sell their hogsheads of Yellow Banks at a loss.[51]

The bubble in western Kentucky's Yellow Banks demonstrated both the promise and the dangers in the specialization between supply and demand. Any success in growing leaf with particular characteristics represented an effort to meet market demand, as the grower in the last chapter asked his merchant in 1846 what color tobacco he wanted the next year. The specific requirements of idiosyncratic buyers meant that tobacco produced to meet demand might not suit anyone else. If a particular buyer or market became glutted with a desirable type, the seller's price would suffer. For the procurer, a need for exclusive characteristics meant relying on specific growers or locations or techniques that met his particular needs. These requirements could easily backfire, as when all available tobacco appeared in the market as "blue as indigo." Nonetheless, such a close relationship between growers and buyers seemed natural as widening supply through westward expansion met widening demand in world markets and the brands produced by American and world manufacturers. The institutional shift of independence had marked the globalization of export markets for American leaf, and westward expansion had meant the development of new leaf types as new environments created distinctions in the "blended mass." Both distinctions were recognized in price-circulars and merchant correspondence. Domestic industrialization therefore accompanied the multiplication of tobacco types.[52]

Old and outmoded institutions continued to act on economic transactions, as did new and changing market relations. Eighteenth-century inspection laws had made distinctions between shipping and manufacturing tobacco. Even when inspection laws changed in 1805, allowing the sale, rather than burning, of refused tobacco, the legal distinctions continued to operate in the merchant characterizations of the leaf, even quite late into the nineteenth century. As one merchant wrote a planter in 1872, his crop (received still in hogsheads) had "plenty of size for shipping, but all too much of a Manufacturing flavor for that purpose." In the nineteenth century, color had become an important signal of quality, although not always an adequate one: the same letter described various hogsheads as "good dark Tobacco in the samples but some brighter and some red without any body and having a dead appearance."[53] In this case, dark tobacco transcended bright. Yet independence and the emergence of a fully fledged domestic manufacturing industry created new institutions, new layers of frameworks for economic activity. New markets and new locations of production enlarged the categories that described the leaf; growers sought to meet the demands of purchasers, while buyers, dealers, and manufacturers solicited specific qualities in the leaf. These desires influenced cultivation methods and the techniques farmers chose in producing commodity leaves.

Categories of commodities enlarged with market demand and westward expansion. Merchants had long described tobacco to its buyers in terms of where it was from, its region of origin. To growers, they described the commodity in terms of what it was for: shipping or manufacturing, Africa or the Southwest. Color had become, in the nineteenth century, a crucial determinant of the purposes of the leaf. Manufactured goods also wore color distinctions: William Sutherlin heard in 1860 from Baltimore "of bright pounds manufactured this season of new Leaf" contrasted with "Dark sweet pounds, 5's, 8's & 10's [and] Dark sweet halfpounds" as signals of what the market desired.[54] Color was not yet sufficient to distinguish tobacco types from one another, but it helped. How color was achieved, however, mattered little to buyers. It marked certain desirable characteristics, but so did region of origin provide signals to buyers, as market purposes did to sellers. An elaborate system had developed in the decades since independence, a commercial system that exchanged the market information concerning raw leaf tobacco and the many products it became. This moment, however, this letter to Sutherlin about his bright pounds, marked the furthest elaboration that nineteenth-century system would take. For that letter Sutherlin received in 1860. A year later, it all fell apart.

CHAPTER THREE

Death and Taxes

Early in 1861, William T. Sutherlin and Peyton Gravely both sat in the Richmond convention and considered Virginia's secession from the Union. From his room in the Exchange Hotel, between Shockoe Bottom and Capitol Square, the commercial and political epicenters of the city, Sutherlin struggled to hold his far-flung businesses together. His commission agents in New York and Charleston sent him missives that offered panicky, unsolicited, and, as could be expected, contradictory counsel about the crisis. He voted against secession on April 4, but the events of the next fortnight changed his mind. Two days after the fall of Fort Sumter, in Charleston Harbor, Lincoln mobilized a militia to quell the insurrection. Virginians then seized the national armory at Harpers Ferry and the Gosport navy yard, joining the Confederacy de facto and officially seceding on April 17, ratifying the action by referendum on May 23. Sutherlin was among the eighty-eight delegates who voted for secession, but Peyton Gravely was not. A rural tobacco manufacturer of an earlier generation, Gravely had been wild for nullification in 1832 but had perhaps mellowed with age. Or perhaps he saw some glimmer of what the consequences of seceding would be.[1]

The Civil War transformed not only the tobacco manufacturing industry but also the entire commodity web that spun out from agricultural leaf production. The emancipation of the slaves shifted economic fundamentals and institutions as well. As laborlords became landlords, so did structural transformations created by the war change the relative costs of land, labor, and capital, the principal factors of production. Military action in the Chesapeake also encroached on the tobacco lands. Battles fought across Virginia and along the entire tobacco coast destroyed the Old Dominion's industrial infrastructure. As a result, both agricultural production and industrial manufacturing took root in new locations, with the usual result: technological changes and new commodities extracted from the plant. In addition, changing institutions—including rebuilt forms of federal taxation—enveloped the tobacco trade as a result of the Union military victory. Taxation transformed the technological systems of agriculture and manufacturing by severing them from one another, distinguishing them firmly from one another, locking the technical processes of harvesting, curing, and preparing for market into the sector called agricultural production. The result was the emergence of tobacco types, regulated into distinctive agricultural commodities for specific purposes, produced in specific locations by means of particular technologies.

The tobacco industry was involved in every phase of the revolt. Cotton grown for export mattered more by 1860, but the older staple still played a role in Southern politics. During secession, at the start of the crisis, tobacco manufacturers joined other Virginians in finally leaving the Union; at the end of the war, retreating Confederate armies set fire to Richmond's tobacco warehouses and burned the city, to stop the Union from taking the Confederate capital as a prize. The industry provided the last capital of the Confederacy, too, when the president fled from Richmond to Danville. Tobacco played a role in every stage of the national tragedy. In the end, new actions of the federal government would transmute the industry into Big Business, its more familiar postbellum form. The predatory, combinatory, capital-intensive tobacco industry of the late nineteenth century marched in step with its age, just as the flexible, ambiguous, modular, and multi-sectoral antebellum tobacco industry had typified its era. This chapter explores the events of the war, the changed political economy wrought by the war, and its impact on tobacco's commodity web. It sets the stage for later chapters that analyze the whole array of changed institutions that emerged from the war, creating a new political economy for industrial production.

—◦ ◦—

Secession severed the commercial relationships that had nourished the infant tobacco industry. Lincoln's election came in the midst of an economic downturn and itself caused an immediate and "consequent derangement of monetary affairs." The presidential inauguration occurred during the first breath of spring, and the blossoming season saw several states secede from the Union to form their own nation. As spring warmed into summer, open warfare between the sections began. Goods, information, money, and credit all faced disruption at the North-South border. Early in the war the Union navy succeeded somewhat in blockading the Confederate coast, segregating the two economies and damaging the South's export relations with world markets. Always at war, the C.S.A. never outgrew its birth pains, the struggle for independence that absorbed all its resources, the losing war that chipped away at its national borders. The resource base of the Confederacy continually shrank, depriving the Confederates of both land and capital, the natural and technological resources to go on. To add insult to injury, Southerners who believed that an agricultural nation would always be able to feed itself found themselves proven wrong. Whatever the South may have been before the war, it was hardly self-sufficient. Southerners needed the outside world. They soon lost it.[2]

The blockade took effect early, in the first weeks of war. It not only interrupted exports but also suspended the coastwise shipping that had served to develop northern markets for southern agricultural and manufacturing sectors. Of course, the blockade was never entire, but the Union navy made every shipping venture riskier. Tobacco firms that had used Yankee merchants to export, sell, or distribute their goods, or dispose of their scrap, sought new arrangements. Peter McEnery, for example, wrote in the first week of May 1861 to London, Liverpool, and Bremen merchants, trying to turn merchant and perform for himself the dealings he had long left to specialist middlemen. "It is a long time since I had the pleasure to address you," he wrote the A&G Maxwell house in Liverpool, before continuing on to state his business: "the object of this is to inform you that I am shipping . . . Tierces & Boxes Manufactured Tobacco, . . . which I will consign to you for sale."[3] As the blockade tightened, fewer and fewer ships made it out of the Confederacy. McEnery's shipment for Europe was seized, "taken from Hampton Roads as a prize," and held for more than a year—during which he wrote constantly to assess its status and his position.[4]

As goods stopped moving into and out of the Confederacy, so did money and credit. In an early example of the difficulties to come, William Sutherlin's brother John took a trip up the East Coast to New York and back down toward home in August 1861. He hoped to "close out some lots of Tobacco" in merchants' hands, pay some debts, and wind up the family's business in Union ports. In Baltimore he discovered that the Bank of Commerce had taken actions to prevent the sale of any of his family's goods, "including Bank Stock, and the Tobacco at the Steam Boat conveyance office and in fact every thing in which you and I were interested in both here and in New York. The Presedent of the Bank of Commerce is a <u>contemptible Black Republican</u> and not disposed to furnish me in any way. [He] declined to discount any paper for me and will take nothing but money in payment of the note." He seemed surprised that secession should have such a result, but how otherwise could he have been treated? Whether traitors to the United States or patriots to a new nation, the Confederates had taken a risk, and their former trading partners now needed to make new calculations.[5]

The first transformation of tobacco during the war was in its location. The old tobacco coast of Virginia rapidly became a battlefield, and Petersburg suffered terribly. Richmond, once nearly the center of the tobacco manufacturing nation, abandoned the industry early in the war and became instead the capital of the Confederacy. In January 1862, when the season for winter work usually lit up with the arrival of spring contracts and manufacturers would customarily sit down to calculate prices and make offers to owners to hire slaves by the year, Richmond's tobacconists sat tight. As spring became summer, and the threat of invasion from the north seemed ever more imminent, the government ordered all "manfacted Tobo to a warehouse . . . to be stored there in order that it may be burned if the chances of the Yankees getting here were at all probably."[6] As the approaching winter brought the anniversary of Lincoln's election, Thomas Branch & Sons, the commercial firm associated with Peter McEnery, saw the writing on the wall. The commission merchant held an auction to dispose of its holdings in manufactured tobacco, all brands and quantities. Not everyone got out so quickly, and much tobacco lay moldering in the city's warehouses. As Peter McEnery wrote to a New Orleans merchant, "no Tobo is being Manufd now in Va."[7]

He was mistaken, of course. The industry that grew like a weed still found its way out of the earth to the sun. Farmers without markets had long turned to manufacturing to preserve their products or add value to them. James Thomas Butler took that route in October 1862, when he invested about a thousand dollars to manufacture his own "Tobacco crop of the year before last." He sold it over the next two years, bringing in about nine thousand dollars altogether. In February Butler and his partner, a state senator, bought manufactured tobacco together and sold what they purchased, doubling their initial investment. In August, with the harvest, Butler began buying raw leaf at a dollar thirty per pound, acquiring 90,000 pounds that by October had advanced to three dollars a pound at market price. Inflation accounts for some of these figures, surely, but nonetheless he received double even that price only three months later. He had begun to refer to tobacco as merchants did, by region of origin and curing method, as "Caroline Sun Cured." In February he hired a Caroline County resident to buy "good sun cured and fine manufactory" leaf for a flat fee per hogshead. Butler was a man who found opportunities in wartime to climb the agricultural-commercial-industrial ladder.[8]

Farther inland, too—and away from the tortured landscape of the U.S.-Confederate border—old-established firms found new opportunities in the dislocations of war. William Sutherlin, for example, found the war an opportunity to expand his business. His correspondents kept close watch on changing markets and demanded specific qualities that consumers liked: in 1862, these were "1/2 pds . . . of a bright mahogany color generally weighing 20 lbs to the caddy (sometimes a little over) in cases 8 caddies to the case." The Fredericksburg merchants "prefer the color for this purpose a richer mahogany. . . . The tobacco should be not liable to mould. We like the weight of the lumps to be about 5 to the lb. We would want the brand <u>red</u> & simply '<u>The Soldier's Comfort</u>,' nothing else." As always, quality and price mattered above all else. "We believe you have our idea about the style color &c. 'a <u>rich red color with some brightness</u>' is about as near as we can describe it. We do not suppose it is necessary to have a very fine wrapper, <u>but one thing is necessary to us; that the price should not be over 17c</u>. . . . We wish you to take your time & put it up well & we will keep you advised when to send it."[9]

The tobacco industry survived the war, and in many of its traditional patterns—but not in all the old places. Not in Richmond. James Thomas Butler's engagement with the industry used the flexibility of the sector in old and familiar ways. Sometimes his business encompassed more and sometimes fewer tasks and transactions, sometimes more and sometimes less of the web that spread from agricultural production through commerce to industrial manufacturing and then through commerce again, until it reached the leaf's ultimate consumer. But Butler kept outside the capital city, where—as if to add insult to injury—the Confederate government used the tobacco industry's infrastructure for its own ends. The political leaders and military authorities of the C.S.A.'s centralized government found that the tobacco economy provided the sorts of large-scale buildings, railroads, and transportation facilities that could splendidly serve military necessity. In Richmond, as in so many of the other industrial cities of the South, the huge tobacco factories and warehouses made useful places for sorting, testing, and storing ordnance, or billeting soldiers, or housing prisoners of war, under infamously horrific conditions. Richmond's indigenous industry collapsed under the city's transformation to a wartime capital. It never entirely recovered.[10]

WARTIME DISLOCATIONS

The war would change everything. Between independence and 1860, as we have seen, Virginia had regularly produced about 40 percent of the total tobacco manufactured in the nation. In 1870 the share was less than 10 percent, and in 1880, just a little more. The value of the tobacco manufactured in the Old Dominion stayed less than 15 percent of the U.S. total for the remainder of the century. New York doubled its share of the total tobacco manufactured between 1860 and 1890, while North Carolina and Kentucky picked up a bit of the business. Virginia would never regain its market share, which scattered and decentralized across the country: no individual state would ever again make more than 30 percent of the total value of tobacco manufactured in the United States (see table 3.1). The industrial structure changed its shape under the stress of military occupation and the rising power of the federal government. The federal government had certainly extended its reach before the war and expanded its grasp in order to meet the demands of the crisis. However, the Yankee Leviathan emerged in its full expression from wartime destruction and national integration. The new power of federal regulation would change the structure of tobacco manufacturing.[11]

Locational shifts also took place in agricultural production. In 1870, the quantity of tobacco grown in Kentucky surpassed for the first time that grown in Virginia, and the center of tobacco production never shifted back east to its origins around the Chesapeake. In 1860, Virginians grew 124 million pounds of tobacco and Kentucky made 108 million pounds; in 1870, Virginia dropped to 37 million while Kentucky's production held nearly steady at 105 million pounds. Kentucky had long stood as the frontier for Virginia's settlers who carried tobacco with them. The cash crop had provided the foothold for western expansion since colonial days. Leaf grown in Kentucky had even developed particular markets as a unique product. The "Western Leaf" served several sorts of distant buyers. Stemming it for export had provided profits in the region since independence. Kentucky's darker tobaccos sold well in Africa, and the specific market qualities desired and packing methods practiced by Kentucky merchant firms verged on manufacturing processes.

TABLE 3.1
The Postbellum Tobacco Industry

	1870			1880			1890			1900		
	Cigars $1,000	Mfd $1,000	% of U.S. total	Cigars $1,000	Mfd $1,000	% of U.S. total	Cigars $1,000	Mfd $1,000	% of U.S. total	Cigars $1,000	Mfd $1,000	% of U.S. total
California	1,910	58	2.7	3,947	10	3.4	3,140	91	1.5	1,888	0	0.7
Connecticut	1,134	15	1.6	787	2	0.7	1,110	0	0.5	1,776	0	0.7
Florida	572	0	0.8	1,348		1.1	8,123	0	3.8	10,891	0	3.9
Illinois	1,348	2,971	6.0	3,765	4,197	6.7	6,942	2,027	4.2	8,741	3,168	4.2
Indiana	794	355	1.6	1,226	2	1.0	1,836	0	0.9	2,537	58	0.9
Kentucky	449	1,648	2.9	983	4,364	4.5	1,058	10,263	5.4	1,507	14,948	7.7
Louisiana	417	162	0.8	507	424	0.8	1,570	282	0.9	1,407	1,084	0.9
Maryland	1,115	657	2.5	1,731	1,531	2.8	2,858	3,216	2.9	2,843	7,054	3.5
Massachusetts	1,598	73	2.3	2,074	218	1.9	4,166	0	2.0	5,298	0	1.9
Michigan	932	1,640	3.6	2,146	1,519	3.1	3,513	4,742	3.9	5,589	3,746	3.3
Missouri	2,084	8,332	14.5	1,524	5,286	5.7	2,155	15,447	8.3	2,746	25,101	9.8
New Jersey	826	459	1.8	1,509	5,064	5.5	1,909	349	1.1	2,648	7,788	3.7
New York	9,265	9,676	26.4	24,768	8,908	28.4	47,423	4,431	24.5	49,028	4,632	19.3
North Carolina	1	718	1.0	46	2,215	1.9	2,552	5,300	3.7	230	13,621	5.2
Ohio	2,767	2,541	7.4	5,019	4,378	7.9	7,025	9,363	7.7	11,240	5,753	7.4
Pennsylvania	5,320	914	8.7	6,907	910	6.6	19,978	3,409	11.1	31,483	1,247	11.8
Tennessee	65	45	0.2	25	260	0.2	237	617	0.4	291	1,541	1.1
Virginia	119	6,936	9.8	484	14,305	12.5	3,728	18,292	10.4	4,844	10,708	7.5
West Virginia	273	5	0.4	453	978	0.4	562	0	0.3	1,060	1,363	0.9
Wisconsin	695	527	1.7	1,347	978	2.0	2,525	1,213	1.8	3,256	1,632	1.7
All other states	2,093	819	3.3	3,385	120	3.0	7,284	3,009	5.0	12,328	1,393	4.0
Product totals	33,777[a]	38,551[a]	100.0	63,980	54,691[a,b]	100.0	129,694	82,053[b]	100.1	161,631[a]	104,837[a,b]	99.9
U.S. total	71,762[a]			118,670			211,747			266,468[a]		

Source: U.S. Census for 1870, 1880, 1890, and 1900.

Note: The values of Cigars, Mfd, and U.S. totals are in uncorrected thousands of dollars. "Mfd" stands for Chewing, Smoking, and Manufactured Tobacco and Snuff; in 1880 (and after), the separate category for "stemming" or "stemmed" tobacco has been added to these. Likewise, in the decades of 1880 and beyond, "Cigars" includes cigarettes and cheroots, when these were counted separately.

[a] This total has been calculated and does not agree with the total presented in the census. "%" has been calculated as a percentage of this calculated total, rather than the one presented in the census.

[b] The data for "Mfd" tobacco include the figures for "Tobacco, Stemming."

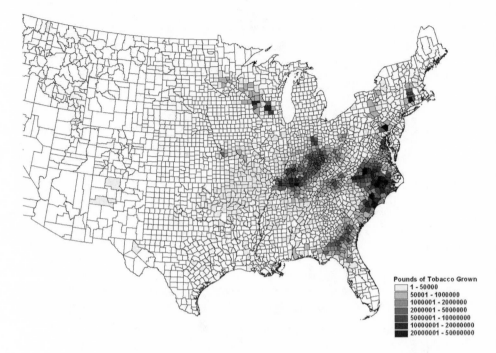

The 1860 census, when compared with the data compiled for 1840 (see p. 41), demonstrated the spread of tobacco agriculture into new regions, as well as its continuing production all across the United States.

Source: U.S. Census for 1860.

Kentucky's Yellow Banks tobacco had been bought by Virginia manufacturers to sell in New York. At the end of the Civil War, Kentucky was ready to take Virginia's tobacco-producing crown.[12]

A new tobacco type emerged in Kentucky during the war—or so says its myth of origin. Two tenant farmers in Brown County, Ohio, ran short of seeds and crossed the river to borrow seeds from a grower in Kentucky. "Although the seedlings grew sturdy and fine-textured, the dirty yellow leaf prompted the tenants to destroy the plants, believing them unhealthy or dwarfed," one source says. The next year a more deliberate borrowing from Kentucky to Ohio led to similar results, but this time the brave tenant "transplanted about a thousand," which, when grown, appeared "healthy and thrifty," with cream-colored stalks. The "freak tobacco" caused a neighborhood sensation. The cured leaf was bright yellow and smoked "bitter" and dry. The new type won several prizes for cutting leaf at the 1867 St. Louis Fair and brought a fine price in the Cincinnati market. The legend had elements familiar from the slave Stephen story of Bright Tobacco's discovery: marketability was a natural quality found in the leaf. So were the methods used to harvest and cure it. "White burley quickly replaced the gummy red burley throughout central Kentucky. It was harvested more quickly—by stalk-cutting—and, as it did not require flue or smoke curing, cured more rapidly. Thus, the tobacco could be brought to market earlier, a decided advantage to cash-starved, debt-ridden Kentucky farmers."[13]

Note the components used to describe the leaf and the reasons for its adoption: the name, the region of origin, and the method of its production. In the account of white burley's acceptance by the farmers of central Kentucky, for example, the harvest method—stalk-cutting—has been described as a property of the plant, and so has the curing method. This jumbling of the qualities of the product (its inherent dryness, as opposed to the "gummy red burley"), along with the methods chosen for its production, had become typical of varietal designations. Once upon a time, the spongy quality of Western Leaf had made it unsuitable for manufacturing: "the quid increases to an unusual size; and . . . loses its blackish rich color, and becomes . . . 'frosted.' This unfavorable mutation is . . . occasioned from the quantity of nitrous fluid it imbibes during the process of vegetation." In the new type, dryness became a desirable characteristic of the region's plant products. Whether or not the characteristics of the commodity were produced by its seed or by the soil or by the techniques used in its production, they were recognizable characteristics that defined the article for its buyers. Remarking the desirable traits helped instruct growers in how to produce it for sale.[14]

If Kentucky found a new bright type for the new era, other regions—some as far afield as Canada—thought to take up some of Virginia's slack. Nearby regions had better luck: south of Virginia, in North Carolina, crops once subsidiary to Virginia's inspection process began to chip away at the Chesapeake's lead. Tobacco moved into new regions, deep into the sandy soils of the Piedmont center of the state, where the coastal plains around the Albemarle Sound began a rolling rise toward the backcountry and the Appalachian Mountains. While the rise in production in the Old North State was never as dramatic as that of the Bluegrass, in 1880 the federal census recorded the emergence of North Carolina's Bright Tobacco type, as "special methods of culture and curing" gave "yellow tobacco . . . year after year its peculiarities." Moreover, the next phase in tobacco's manufacturing sector would center on the town of Durham Station, which would grow up with the new tobacco industry and become "the tobacco center of the world," as Richmond once had been. As the South's bid for independence failed, her capital on the Chesapeake's tobacco coast entered history as a myth of agricultural production and plantation paradise.[15]

The last days of the Confederacy took shape as part of the end of Virginia's industrial dominance. Although surrounded by federal forces, the Confederate Congress in Richmond seemed unprepared for surrender when it adjourned as scheduled in the middle of March. On Sunday, April 2, during churchgoing morning hours, word arrived that Petersburg had fallen to the North. The Confederacy's president, Jefferson Davis, prepared to evacuate his cabinet to Danville. On the presidential train rode the last of "the archives and treasure of the Confederacy," the latter consisting of gold coins and ingots in legendary quantities. The president left Richmond shortly after midnight, on one of the slow trains of the late Confederacy. Out of the five railroads that had served the city at the start of the war, by spring 1865 the Richmond & Danville was the only functioning line still in Confederate hands, so Davis headed down toward the river Dan, to the Virginia–North Carolina border, heading for the home of his compatriot and quartermaster, the tobacco manufacturer William Sutherlin, in Danville.[16]

When the Confederate government fled Richmond for Danville, it torched the city behind it. The soldiers began the conflagration by setting fire to the city's tobacco warehouses. In keeping with the standing order that accompanied most Confederate retreats, the goal was to burn the great Southern staples that might otherwise fall into enemy hands. Very little cotton but considerable raw tobacco lay in the warehouses that lined the waterfront and dotted the commercial parts of the city. Shockoe Warehouse was among the first to go—soldiers set torch to it and a few others and moved on to the bridges. The wind that picked up from the south did the rest. It blew over the river, fanning the flames deep into the commercial center of town. The city burned for days and smoldered for weeks. The industry that centered there provided the kindling that fed the fire. That industrial structure incorporated the labor arrangements that drove the nation apart, which had changed the city into the capital of a belligerent nation and then destroyed the city, the industry, and the Confederate nation. The giant inferno of leaf in which the city burned reflected the fiery end to which the economic system had brought the region.[17]

WHEN THE WAR WAS OVER

The full extent of the wreckage would not be assessed for more than a century. At a certain level, it is still being calculated today. The war's long-term effects reverberated for a long enough period to make calculating its impact the work of historians rather than contemporaries. Those who have done the math tell us the war cost $6.6 billion in direct costs: government expenditures, physical destruction, and loss of human capital. Indirect costs, including the decline in consumption and the effects of cotton prices and emancipation, add another $3.7 billion—although this is more tentative a figure. About half a million young men died. The war destroyed the value of investments in slaves, in land, and in capital. It retarded the expansion of the antebellum networks within which information, finance and credit, and commodities flowed. Yet, as Charles and Mary Beard recognized at the start of this century, political institutions and private economic organizations developed during the war in ways that shaped the later triumph of Big Business and the corporate form. The flowering of federal power nurtured the economic growth witnessed in the postbellum decades and at the start of the twentieth century.[18]

Success stories did not immediately emerge from the ruins. The reports on creditworthiness made by R. G. Dun and Company (with head offices located in New York and on-the-ground reporters in the locales where commerce operated) reveal a great deal about the impact of the war on Southern business. That effect was uneven, and the reports capture the uncertainties the war left in its train. The Dun records began describing Southern business in the 1840s and continued until the 1870s and in some cases the 1880s. None were made during wartime. The interruption of the war to these reports captures in stark contrast the distinction between the antebellum and postbellum worlds. When the Richmond reporters began making their first comments on creditworthiness after the conflagration, they marked many firms as simply out of business. Little is known of the myriad tobacco manufacturers who failed in the war. William Greanor, for example, ran a large factory with his son before the war and made a fortune from it. In January 1861 he was a "Rich man" worth $200,000 or $300,000. In May 1865 all that could be said was "Out of bus[iness]."[19]

For manufacturers familiar from the last chapter, those who worked the antebellum industry, the effects of the conflict differed. Peter McEnery, for example: his Petersburg partner of the 1850s, James M. McCulloch (makers of the Mac & Mac brand), did not survive the war. The various firms of Thomas

Branch & Co., the merchant firm of which Peter McEnery was first a subsidiary, did endure, but only by splitting into specialized pieces: the parts of the enterprise located in Petersburg remained commission merchants, while the Richmond branches turned to banking and investment. By the 1880s the balance had tilted almost entirely into finance, where it remained for the life of the firm, only expiring in 1976. As for Robert Leslie, one of the earliest movers from factor to factory, his papers at the Library of Virginia end in 1859, although he remained alive, according to some accounts, until 1879. The Gravely family survived and continued manufacturing on the North Carolina–Virginia border, but its travails provide a marker for the transformations affecting the industry after the war, as we shall see.[20]

Some tobacco manufacturers did extraordinarily well out of the war. After all, as a merchant wrote one manufacturer during the winter after Lincoln's election, months in which the nation girded for civil war, "people will Chew, Come what may." Sometimes the credit reporters could depict successful business dealings out of war's dislocations. The giant ledgers of R. G. Dun and Company recorded many a wartime success. See, for example, Robert A. Jenkins of Williamsboro, North Carolina, who did a flexible antebellum tobacco business of "farming & manufact'g Tobacco—trades a good deal." His five or eight thousand dollars worth of land and slaves in 1859 had become, by 1867, twenty thousand dollars. Four years later the figure was thirty thousand dollars. William Sutherlin also made himself another fortune. In the hungry months and years immediately after surrender, while his friends and neighbors struggled along with so little, writing him to beg for a few pitiful dollars, Sutherlin was satisfying his pent-up consumer hunger by ordering a pair of "Walnut Serpentine Bedsteads" from a New York firm and fancy liquor such as champagne and sherry that he must have missed during the blockade and while out of touch with his New York merchants.[21]

Business relationships slowly, tentatively re-formed across the North-South border, trying to put the disagreement and violence behind them and find their correspondents, inquiring after their doings, commercial and familiar. The secretary of the Chicago Stock Exchange, for example, wrote an old friend in Richmond in April 1865 to learn how his friends "have survived the trials of the past 4 years, although we may differ in political sentiment yet as individuals & families we may yet be friends." He wanted to know "what the result has been upon your personal affairs of course we are not able to determine but fear that it can hardly have otherwise than disastrous," inviting him north: "if you are at all disposed to seek other fields for business than the south let me solicit you to come to our City_ it has been steadily prospering for years & more so for the past 3 & fields for men of moderate means & large energies are always open." If some people harbored grudges, business seems to have put the conflict behind it, ready to move forward, as business does, seeking profits even in the remnants of tragedy.[22]

And yet, as the rebel states straggled back into the Union, they found it modernized. As the political scientist Richard Franklin Bensel has argued, "the American state . . . emerged from the wreckage of the Civil War." But not by chance: "the very process of secession, war, and reunification both strengthened the American state in every dimension of institutional design and sub-stantive policy and committed the entire apparatus to the promotion of northern industrial development and western settlement." Some historians find a longer timeline, but the Civil War was the United States' war of national integration, akin to so many other nineteenth-century European and imperial wars. Wartime legislation and new bureaucratic structures gave the federal government fresh authority over business practices and market processes. For Southern industry, the return to the Union meant a return to a federal government armed with new policies and consolidated power.[23]

One example of the federal government's new position in economic activity was taxation. During the war, new internal revenue laws established federal taxes on many industries "to meet in part the exigencies" of war. Some lasted past Confederate surrender. Designed to pay off the vast debt incurred during the war, and later to pay out veterans' pensions, taxation furthered the federal government's resources to exercise its postbellum powers. Since independence, the nation's leaders had worked within the taxation provisions of the Constitution to develop "the central instruments of government," especially financing its activities. Property taxes financed local and state governments, while tariffs provided "the core of federal finance." New taxes applied in order to address "the first great national emergency," the Civil War, included income tax, which fell on only the richest; as one Congressmen constructed it, the law could not "allow a man, a millionaire, who has put his entire property into stock [to] be exempt from taxation, while a farmer who lives by his side must pay a tax." In 1872, at the behest of the richest citizens, Congress let the income tax pass away. Excise taxes on liquor and tobacco, which provided about half the federal revenue in the mid-1890s, remained.[24]

REVENUERS

Civil War–era taxation transformed the tobacco industry. Protested vigorously at every point, adjusted and readjusted in the details of its provisions, taxes on manufactured goods took more than a generation to work through the entire commodity web—forward, into consumption habits, and backward, into agricultural production. Assessed in order to pay off the federal debt incurred for the war, excise duties bore down on the South after Confederate defeat. The effects of these imposts on tobacco resembled those on another large Southern business, liquor production. Taxation helped shift the manufacturing tasks of moonshine distilling into the modern liquor industry. "[C]aught between the giants and the five-gallon-still moonshiners," legitimate, tax-paying, small-scale producers cooperated with the collectors in prosecuting the competition. Consolidation swept through the industry. Likewise, taxation pushed the tobacco industry toward combination. Federal taxation of the Civil War era did not alone create the trend to Big Business that gathered steam at the turn of the last century, but its impact on these Southern industries demonstrates that government policy, as well as the invisible hand of markets and the visible hand of managerial control, contributed to the great merger movement and the conglomerate tendencies of the age.[25]

The internal revenue law passed on August 5, 1861, went into effect at the start of the next fiscal year, on July 1, 1862. It was tweaked, adjusted, and amended every few years thereafter. The revenue commissioner might protest that even the "smallest change will for a time work inequalities," and that "[a]lterations even in the machinery of the law are always attended with embarrassments," but Congress sometimes amended its earlier actions only a few months later. For example, cigars shifted throughout the 1860s between uniform rates and taxes graduated *ad valorem* (according to their value). The tax laws made its distinctions among the grades and types of manufactured tobacco products along three axes: the intended purpose or mode of ingestion, the method of manufacturing, and the value of the raw materials. Smoking tobacco began at five cents a pound, which rose to twenty-five cents in June 1864. A year later the taxes levied on better smoking tobaccos rose to thirty-five cents per pound, but "that made exclusively of stems" stayed at a quarter dollar. Chewing tobacco fell into entirely separate classifications: fine-cut chew and plug began by paying fifteen cents per pound but rose to thirty-five cents in June 1864, and then forty.[26]

The Southern branches of the tobacco industry would not feel taxation's effects until after the Confederacy's surrender, but the Union's industry felt the impact earlier. Three thousand cigar makers in New York City were thrown out of work, according to one wartime protest, "and a corresponding number of packers and strippers [were] rendered almost homeless through the operation of the act." The cigar makers had at least the benefit of tariffs, enacted in 1861, 1862, and 1864. According to their proponents, these counteracted the taxes and worked "against the importation of cheap cigars from Germany" to give the industry "an impetus" to "continue in practically uninterrupted growth." Such tariffs provided little protection for the industry more typical of the South, however. Manufactured chewing and smoking tobaccos faced little foreign competition: plug (the name given in the tax laws for all manufactured tobacco) was distinctly a homegrown industry. The Kentucky legislature began protesting the taxes before the bill even went into effect, arguing that the law would fall "with great severity" on the regions of the state already most distressed by the war. Kentucky's lawmakers instead pushed the idea of *ad valorem* taxation, "owing to the great difference in the value of different qualities and grades of tobacco."[27]

The New York cigar manufacturers claimed, as others would later claim, that the solution was to throw the tax onto other sectors of the industry. In the most stark statement, cigar makers argued that "the only proper and just mode of taxing tobacco is to tax it in the leaf." The federal government, however, declined to limit its taxes to agricultural products, for obvious reasons. The romantic view of farming that prevails in American political culture likely made even a wartime, Republican, pro-business administration reluctant to lay a direct tax on farmers, especially when the manufacturers added so much of the value to the final good. Farmers were adept at complaining about laws they found "unequal, unjust, and oppressive to the agricultural interest." More to the point, even an urban or industrial legislator could understand that it was unwise for the nation's export commodities to bear the entire burden of federal taxation. That would raise their prices and curtail their buyers in world markets. So, if not taxed at the barn, where and when should the tax on tobacco be levied? The government wanted revenue and therefore established a tax on this commodity. But how to define the article being taxed? When to tax tobacco? At what point in its manufacture?[28]

It took years of fine-tuning to decide how to assess the tax. The earliest federal tobacco revenue laws, written during the first months of war, attempted to define the product by its market purpose, by the processes that produced it and the value of its raw materials. This worked fine for cigars.

A cigar is just a cigar, an individually manufactured product, countable one by one and taxable in the thousands. But manufactured tobacco—the kind typical of the Southern industry, sold to consumers to smoke or chew, as they pleased—raised a whole host of issues likely unimagined by Union legislators. Tinkering with the rates and their assessment did not solve the most basic problem of classification: thing uncertainty, or how to distinguish one type of good from the others. The tobacco products of the Southern industry were as hard to pin down as its sectors and firms. Consumers blurred the boundaries among methods of ingestion, and dealers and manufacturers utilized similar flexibility to increase their profits. The revenue laws conflated the value of raw materials with the part of the leaf used, strips or stems, and so its definitions rested on modes of manufacturing. And only the manufacturer could know for sure what manufacturing techniques he employed and what materials he fed into them.

Take the revenue law of July 20, 1868, one of the federal government's more ambitious attempts to make the law conform better to practice. It simplified categories by taxing manufactured tobacco products at only two rates. All chewing tobacco paid thirty-two cents to the government, as did the best smoking tobacco. Yet earlier statutes had singled out smoking tobacco "made exclusively of stems" for a lower rate of taxation, and the 1868 law preserved that distinction but extended it to smoking tobacco made "of leaf with all the stems in, and so sold, the leaf not having been previously stripped, butted, or rolled, and from which no part of the stems have been separated, by sifting, stripping, dressing, or in any other manner, either before, during, or after the process of manufacturing." Such careful descriptions did not forestall fraud; they made it easier to accomplish. The problem lay in the very act of definition, of what processes made what products. Earlier periods witnessed similar difficulties. This should be familiar. Colonial inspection laws—by specifying the characteristics that made leaf merchantable—had solidified the cultivation techniques used to produce the crop. Federal definitions of types of manufactured tobacco products worked in comparable ways.[29]

Low-grade smoking tobacco made of unstemmed leaf paid only sixteen cents a pound in tax, while even the "cheapest grades of plug tobacco" paid thirty-two cents per pound in tax. Since manufactured tobacco could be consumed in any number of ways, its makers evaded the law by sweetening the plug for chewing but paying taxes only on unstemmed smoking tobacco. Indeed, "the poorer classes of consumers" may have used sweetened plug for smoking, yet the customer's price for either was about the same, despite the difference in tax rates. Manufacturers likely pocketed the difference. The internal revenue commissioner complained that such manufacturers "allege that they do not know, and are not bound to know for what purpose their goods are bought and used." In fact, after only a few years of operation, the revenue office recognized that "[t]o make the rate of tax depend on the process of manufacture unquestionably opens a wide door for fraud. No one can determine by inspection of the product whether a given sample . . . contains all, or more, or less than the natural quantity of stems." Was it stemmed, and was it sweetened, and how would it be used? Few could tell, although tax rates relied on such information.[30]

Tax assessors could not judge how much stem was in a cut tobacco prepared for smoking, nor if it had been sweetened, nor whether it was chewing or smoking tobacco. In other words, government officials had no idea what the tobacco product they saw really was and, consequently, what tax was due. Their use of the term "plug" to describe all manufactured tobacco reflects their difficulty. The modular quality of the manufacturing processes made it difficult for any law to pin down what value had been added at any one point. If a manufacturer could sweeten the leaf and, as the internal revenue commission complained, still classify the consumer good as an agricultural product (and therefore a raw material) by leaving the stem in, then it was not a manufactured article subject to taxation. Leaf dealers took advantage of these laws to sell raw leaf to consumers, paying no taxes because their "process of preparation involv[ed] neither the use of any machine or instrument, nor any process of pressing or sweetening." Thus, leaf dealers undercut the cost of manufactured (and tax-paid) tobacco, or profited from a higher price designed to incorporate the taxes they did not pay.[31]

The internal revenue commission found that the solution lay in shifting the criteria that defined a manufactured product. Rather than characterize manufacturing as particular processes, or according to the finished article that those methods made, the changing revenue laws crept toward defining the product by who bought it. If a consumer bought tobacco, it had been manufactured. If it entered a manufacturer's hands, it was a raw material. All the intricate distinctions between manufacturing methods made a taxable unit called plug, or called cigars. The transition of definition began with the act Congress passed on June 6, 1872. This law shifted the machinery of taxation and thereby changed the definitions under which it operated. The government simply charged a uniform rate for all manufactured tobacco and sold licenses to perform certain transactions, but not others. The activities that had distinguished the sectors from one another—the tasks that dealers could do but manufacturers could not, that farmers could do and peddlers could not—were no longer regulated. Anyone could now legally sweeten or stem a leaf. The revenue laws would dictate industrial structure rather than processes: from whom a license holder could buy, and to whom one could sell. Dealers in raw leaf could not sell to consumers. Licensed manufacturers were the only ones allowed to do that.[32]

To demonstrate compliance, dealers and manufacturers had to keep precise records. They had to measure the commodity at every stage of its processing and document carefully all purchases and sales. License holders could sell only to other license holders in an order designated by the federal government. As the internal revenue commissioner described the law, any leaf dealer who sold tobacco to "an unauthorized manufacturer" was liable for fines and penalties. The regulations extended all the way down to the knapsacks some traveling peddlers carried. To make taxation effective and fair, its classification of the industry's organization had to span the entire commodity web, extending even to people who sold consumer goods out of their wagons. Otherwise, that one weak link in the chain could affect the industry's whole structure, creating a black market of dealers and manufacturers who supplied untaxed tobacco to the retailer. The old flexibility of the trade, the blurry borders between growing, marketing, and manufacturing, made tobacconists of any sector able to seek any path around the law. In practice, of course, even this system did not work that smoothly. For one thing, the leaf defies consistency in measurement. As the antebellum industry knew, the leaf gained and lost weight at stages in its processing that never followed entirely predictable patterns.[33]

To add to the difficulty, the revenue system hardly functioned with machinelike regularity. The constantly changing laws created confusion even for government officials: "Months are required by the revenue officers, especially those remote from the central office, for learning the new requirements of a statute . . . it cannot be expected that those whose attention is not devoted to its study and administration should earlier ascertain all that may be required of them." Some revenue officers were dishonest, or simply greedy, and the laws—with their precise definitions of commodities and careful demarcations of industrial sectors—provided ample room for official manipulation. Two decades after the internal revenue laws first went into effect, public opinion, newspapers, and state governments were still able to lay "many and serious charges of misconduct" against the revenue officers. State legislatures in tobacco country protested on behalf of their citizens "being prosecuted in the Federal court for the most trifling offenses." Dishonesty lay in the process of "enforcement . . . engendering strife and confusion among the people." As one observer of the industry put it, "it is a common thing for the Government to close Tobacco Factories and Manufacturers in these parts, the Tyrany of the Revenue men makes dealers in Tobacco more or less tricky."[34]

Revenuers seized tobacco factories right and left. The size of the firm, the value of its products or wealth of its proprietor—these made little difference to the excise officers but certainly helped a tobacconist weather a revenue case. Take D. L. Dyson, a North Carolinian who had "engaged in farming" before the war began, "selling goods in a small way & manfg tobacco." His "Factory" was "seized for alleged violation of the Internal Revenue laws" in June 1874. He fought for three years before he "compromised & pd in part. but the suit or libel vs. the Factory is still pending." The government threatened to take his homestead but later abandoned that part of the case. His situation became ever "more hopeless," and five years after he first ran afoul of the law he still "had not emerged from . . . under the Harrow." On the other hand, substantial resources made for better luck: J. B. Pace of Richmond was arrested "for defrauding the govt" and had his "Factory seized" early in 1870; "he went to Washington and stated he had taken advantage of Revenue Laws" and offered to pay twenty thousand dollars. "[H]e offered to make good they accepted his offer." He had "settled all his difficulties with the Govt" within six months.[35]

The problem lay far deeper than corruption or dishonesty. Even the most honest system of collection staffed by virtuous civil servants would struggle with the difficulties of defining the commodity precisely at every stage of its manipulation. Asymmetric information was one expression of this difficulty: manufacturers knew what had been done to produce a product, but revenuers did not. The large economic literature on quality uncertainty began by considering the market for used cars. The analysis demonstrated the inefficiency of markets at setting prices in conditions of information asymmetry: the seller knows how good or bad her car is, but the buyer has no way of knowing for sure. Information, however, is just another word for knowledge, and the difficulties inherent in recognizing when a tobacco product was manufactured, as opposed to simply prepared for marketing as an agricultural good, presented a real problem for revenuers. In such a case of thing uncertainty (when the question was not merely quality but also type—as if the difference between a Ford and a Toyota were also unknown), negotiations to determine the worth of the thing could not rely on what it actually was. The government solved the problem by licensing each link on the commodity chain. This shifted the uncertainty away from the article itself and resolved it into a legislative framework that classified buyers and sellers instead of goods.[36]

The net effect of taxation was to separate the sectors so as to define manufacturing as an activity distinct from agriculture and commerce, one whose products brought income into the government's coffers. The tobacco industry had taken root in the interpretive flexibility of the commodity, the blur between boundaries of the firm and the sector, between agricultural marketing and manufacturing processes. Taxation terminated that flexibility and defined manufacturing as something distinct from the agricultural technologies and commercial activities within which it had sprouted. Take the case of Thomas J. Noble of Virginia, a Confederate veteran who set up shop as a "tobacco prizer" after the war. Apparently his products came close enough to final consumer-ready goods to classify him as a manufacturer, for he experienced "some trouble with the Revenue Officers in regard to the Stamping of some Tobacco" that made him very "hard up for money" by 1870 and drove him into bankruptcy a year later. In earlier generations, planters and merchants such as the Gravely family might use price signals to choose whether purchasing their neighbor's crops was speculation or raw material acquisition for manufacturing. Now that path into the rural manufacturing sector was overgrown so thick with laws that only those with clear intentions and license fees could follow it.[37]

The difficulties of producing under such careful delineations drove small producers out of business and made entry into manufacturing more difficult. Men on the margins could not comply with the new regulations. As Erastus Mitchell of North Carolina discovered, the threats of internal revenue action jeopardized even a successful business. This tobacco dealer and sometime manufacturer found his credit worth downgraded in the new regulatory environment, as the Dun reporters concluded that "he may be d[oin]g well but a seizure may ruin him." Whether honest or not, charging fees for manufacturing licenses meant that a leaf dealer could not choose manufacturing to preserve his investment, or to increase his profits, if an expected rise in the price of raw leaf failed to materialize. Manufacturing required more effort, and a license necessitated consistent effort over the period the permission covered. Rather than a task anyone could take up, an extension of agricultural marketing techniques and structures, tobacco manufacturing became a business regulated in such a way as to bring revenue into the federal gov-

ernment. Taxation therefore contributed to industrial consolidation. The tobacco manufacturers' combination that drew federal antitrust prosecution in
the twentieth century at least partly emerged from these distinctions created
by government policy.[38]

MONOPOLISTIC TENDENCIES

The shifts in the industrial structure created a shakeout, in which large and
profitable firms bought smaller ones, and marginal establishments went under
the harrow. Take the case of John F. Allen, the son of the cigar maker who had
started his career in antebellum Richmond. He was the overseer in William
Barret's factory producing Negro Head, the mixture famous as a brand among
British consumers. He had taken risks and formed multiple partnerships that
eventually moved him from managing someone else's factories and slaves into
his own profitable manufacturing enterprise. His business had finally turned
the corner when he joined forces with a factor who bought widely in the public warehouses, for the European markets, which increased his access to the
leaf. After the war, Allen followed the trend toward enlargement that taxation
initiated. His new connection with Lewis Ginter was a highly profitable one,
earning the two men immense profits and allowing Allen's retirement from
the tobacco business in 1883. The new firm still bore Allen's name, although
only Ginter—and his new partner John Pope—stood at the helm. When the
American Tobacco Company formed its combination, Allen & Ginter was one
of its five constituent firms.[39]

The American Tobacco Company or ATC—famed combination, monopoly,
or trust—resulted from the taxation policies of the wartime federal government and the end of flexibility among tobacco products. The legendary rise of
the ATC trust is an oft-told tale, one that fits neatly into the stories historians
tell about the Gilded Age, the triumphs or depredations (depending on one's
preferences) of unchecked corporate industry, and the eventual response of
the federal government. The colorful president of the ATC, the raw-boned,
red-headed North Carolina boy James Buchanan "Buck" Duke, became the
Southern version of the robber barons who dominated Big Business at the
turn of the last century. When the 1900 Census of Manufactures dismissed
the early tobacco industry as crude hand manufacturing, unworthy of analysis, it set the stage for this simplified story. Those who consider this period as
the entire history of the tobacco industry miss the more compelling story of
structural change: how a complex commodity web gave way to a simple com-

modity chain as a result of the shifting political economy. Taxation marked one element of the transition away from the antebellum tobacco industry, which had flourished in the business structures of plantation production, toward the cigarette century and consumer loyalty to mass-produced brands, eventually Marlboros and Camels.

Traditional history tells us that three things contributed to the emergence of Big Tobacco: the entrepreneurial energies of Buck Duke, the development and adoption of the Bonsack cigarette-rolling machine (patented in 1881, adopted by the ATC in 1884), and the cigarettes it rolled that met the burgeoning demands of a new consumer age. Entrepreneurship certainly played a role in the rise of the ATC. Washington Duke, yeoman farmer and family patriarch, started the business from the leaf scraps he found on his farm after walking home from Confederate service. That he fought to defend the slave-holding South has not tarnished his image as a subsistence farmer. That he sought credit in New York in 1856 might complicate the story, however. Yet it was his son Buck who propelled the family business into oligopsonistic enterprise. Still, while students and scholars alike celebrate industrialists like the Dukes, the more anonymous managerial class, the "hierarchy of salaried executives," better represents the organizational changes of the age. Alfred Chandler called it the "visible hand," when management decisions and allocations replaced the invisible hand of market forces. While most managers made no great name, still individuals can demonstrate what business history adds to the story—the entrepreneurial spark of individual initiative, the pluck and luck and character without which economic change and growth never happen.[40]

Adding individuals into the history of economic structures provides modern scholars with the continuities that accompany structural shifts. Even when unnamed, humans who survive transformations can illuminate them, as in Gavin Wright's analysis of the postbellum agricultural sector. Individual planters who persisted through the war and the liberation of their slaves changed their acquisitive strategies—had to change—in the new political economy created by emancipation. No longer owning humans, planters changed from laborlords to landlords. A similar role for individuals can be seen in industry, as in the example of John F. Allen. He retired from the business, but his firm survived and became part of the ATC, the combination that most successfully used the monopolistic strategies molded by taxation. The persistence of individual firms and entrepreneurs demonstrates that the ATC trust was not entirely distinct from the earlier tobacco industry, although business tactics took novel forms in the new era. There was overlap between the two periods, as economic actors, both people and firms, struggled to make money in the changed environment. Institutional change has far-reaching, sometimes slow effects. People can help scholars mark the transition from one industrial structure to the next.

The second part of the traditional history of the ATC, however, sounds suspiciously of technological determinism: the Bonsack cigarette-rolling machine caused the monopoly, as machines allowed certain producers of common and indistinguishable goods (nails, soap) to achieve economies of scale that eliminated their competition. The best treatments recognize that technology is just one element of economic structure, as regulation is another. Naomi Lamoreaux argues, for example, that firms with the most to gain possessed "a large potential market" and utilized "an innovation or discovery (or change in the tariff) . . . to exploit that market." Yet other scholars of American industrialization have been known to enthuse that "the new mechanical devices that issued from the nation's workshops contributed greatly to increased efficiency and production," further describing Bonsack's invention as "truly spectacular." In this version, Buck Duke's purchase of Bonsack machines "allowed him immediately to increase tenfold the daily production of cigarettes." This leap in production efficiency, however, actually "saturated" the "small market for cigarettes." Not many people used the paper-wrapped cigar. "To us the smoking of cigarettes savours of the effeminate, and is not suited to the English nation," according to a London trade periodical. Americans mostly chewed, smoked cigars, or smoked pipes.[41]

Finally, however, cigarette smoking did become the dominant mode of ingestion—although not until around World War I, three decades after the development of the Bonsack machine. Moreover, there were several cigarette-rolling machines to compete with the Bonsack, a case of "multiple, simultaneous invention." The French regie used Decouflé machines. These so threatened the Bonsack model that Buck Duke bought the U.S. rights to the Decouflé devices just to prevent his competitors from adopting them. In England, John Player & Sons used a mechanism called the "Elliot." Historians of technology use systems theory to grasp the way a machine requires adjunct technologies to make it work, as lightbulbs relied on hydroelectric dams and automobiles needed pavement. Cigarette making became effective with the addition of machines that wrapped the cigarettes into packs printed with identifiable brands and attention-getting designs. At the other end of the production cycle, redrying machines controlled the leaf's moisture before the to-

bacco entered the Bonsack machine and also helped the ATC in the organization of its purchasing. Constant innovation in technology marked the ATC's
business and stretched well beyond the Bonsack machine, reaching into both
raw material acquisition and distribution.[42]

Since none of these three elements of the traditional story stand up to
analysis, we must ask: how did the ATC achieve its dominance of the tobacco
industry? In 1890 the five largest manufacturers of cigarettes formed the
ATC, with Buck Duke presiding. From the start, the ATC controlled 90 percent of the country's cigarette business. Since so few Americans consumed
their tobacco in cigarette form, however, the firm needed traction in the market for manufactured tobacco (called plug by the government). In 1881 the
demand for plug was two and a half times that for smoking tobacco in any
form. So the ATC bought a plug manufacturer, along with its investments in
snuff and smoking tobacco firms, and the nation's principal cheroot maker.
Between 1894 and 1897 it battled for control of the plug market but lost. The
company chose "Battle Ax" as its weapon and set the wholesale price of this
brand of plug below the cost of production. It used profits from the cigarette
business to subsidize these sales, hoping to destroy its competitors in price
wars. The company spent millions but, in the end, controlled only one-fifth
of the nation's plug-producing capacity. Not until 1899—nearly a decade after its formation—did the "monopoly" see its share of all the plug and twist
produced in the United States exceed one-quarter.[43]

After all, other firms were also seeking to corner the assorted markets that
together composed the tobacco industry. This makes sense. Since the impulse
toward consolidation had roots in the regulatory environment, more than
one firm felt that influence. Plug manufacturers had already banded together
before the ATC did the same for the largest cigarette producers. Tobacconists
had formed the "Trade-Mark Protective Association of Plug-Tobacco Manufacturers" in 1884; by 1890 there existed also an organization of plug makers known as the Manufacturers' and Buyers' Association, which "apparently
had no monopolistic purpose," instead intending to prevent price cutting by
wholesalers. Already, plug manufacturing was "concentrated in the hands of
comparatively few concerns," but this did not make the ATC's goals easier to
reach. In fact, there were two kinds of plug manufactured from two of the
most important leaf-growing regions of the United States. Flat plug originated in the Southeast, where Buck Duke had grown up, and was made principally from Virginia and North Carolina leaf. On the other hand, Kentucky
plug factories used mostly burley tobacco. Since the two types of plug used

different raw materials and served specific market demands, price wars were largely an ineffective tool in the ATC arsenal.[44]

Therefore, the ATC, after its initial disappointment, pursued a new strategy in plug production. It abandoned the flat plug produced on the East Coast into the hands of another monopolist, R. J. Reynolds—the scion of the antebellum Patrick County manufacturers Hardin W. Reynolds and David Harbour Reynolds, who bartered goods for rustic produce on the Georgia frontier in the 1830s. Instead, the ATC founded the Continental Tobacco Company to consolidate in one of its branches the production of the "navy" plug and thus "confined [itself] to the manufacture of tobacco from the burley leaf." Then, the next year, the ATC bought the firms that made "the materials and supplies used in its factories, acquiring for this purpose interests in a bagmaking factory and a licorice business." Acquiring licorice interests proved a smart move; the white burley that had emerged during the Civil War apparently shared the "spongy property" that had characterized Kentucky leaf in 1842. It absorbed more "sugar, licorice, and other casing materials" than did the Chesapeake leaf—nearly 100 percent of the leaf content, by weight. The ATC put competitors at a disadvantage by the prices they charged for licorice, which every sweet navy plug needed. For the plug made of Western leaf, control of these raw materials (licorice and packaging) contributed to the combination's eventual success in mastering the multiple branches of the tobacco industry.[45]

In the end, however, to control the market, the ATC changed the consuming habits of the public, partly by advertising. The Allen & Ginter firm had pioneered in this realm, enclosing cards in cigarette packs to stiffen the wrapping. The ATC printed these inserts with pictures of scantily clad actresses (among other subjects), images that ran in sets and thereby inspired repeat purchases of that brand. This moved consumers into brand loyalty and cigarette consumption: if the industry in 1881 made 250 percent more plug than smoking tobacco, by 1897 it was only 100 percent more. A twentieth-century investigator then reported that "plug tobacco has lost ground, while the sale of smoking tobacco has increased rapidly." By 1905 plug output was 172 million pounds, compared to 165 million pounds of smoking tobacco. The gap had narrowed to almost nothing, and cigarettes finally overtook plug consumption in the 1910s. Consumer loyalty pressured retailers to stock specific brands. The ATC first sold its goods on consignment, then switched to owning its distributors, and later, under legislative pressure, instead gave rebates to retailers who sold at ATC prices. Each of these strategies helped end old patterns in which merchants selected goods and set their prices. Far more than production efficiency, controlling distribution led to the ATC's market share, which reached 70 percent of the entire industry—all tobacco products—during the cigarette century.[46]

Shifts in the regulatory environment only cemented the trust's power. The Spanish-American War in 1898 motivated the government to raise taxes on tobacco manufacturers, and retail prices went up accordingly. Three years later the tax dropped "to its former basis, but the Combination was powerful enough to keep its prices on the higher level. It thus absorbed practically all the benefit of the reduction." According to the federal antitrust researchers, this demonstrated the danger of "monopolistic conditions not fully recognized. The tax reduction, of course, was intended to benefit the consumer. As a matter of fact, it benefited almost solely the controlling interest in the industry." Yet the pattern was set. In the twentieth century, the combination had so overpowered the industry that it cleared profits of nearly 70 to 80 percent on "the entire plug, smoking, fine-cut, cigarette, and little-cigar business of the country." As the list itself demonstrates, the company's profits were divided among the different branches of the industry—and its control of each branch was lodged in a separate combination of firms. Its subsidiaries included the American Snuff Company and the American Cigar Company, while the Continental Tobacco Company combined the plug-tobacco producers.[47]

Technological change was not the cause of consolidation: the whole industry was already increasing the production of cigarettes *before* the ATC installed the Bonsack cigarette-rolling machines in 1884 (see table 3.2). Mechanizing production helped tobacconists meet their goals, but their efforts predated the introduction of machines. Even after the ATC adopted the machines, it took several more years for the ATC to get the machines working successfully. The changing political economy better accounts for the changing product mix of tobacco manufacturers than does the adoption of the new machine. Moreover, the consumption of cigarettes did not overtake the consumption of other manufactured tobacco products until World War I. Instead, a more satisfactory explanation for the rise of the cigarette century can be found in the institutions that shaped business strategies and the adoption of new technologies. The political economy that emerged from the Civil War, with enlarged federal powers and new tax policies, better accounts for the goals that technological change merely helped to accomplish. This becomes obvious when we recognize that many firms made efforts similar to those of the ATC.

TABLE 3.2
Cigarette Production in the United States

Calendar year	Number of cigarettes and "little cigars" (in millions)	Calendar year	Number of cigarettes and "little cigars" (in millions)
1880	533	1894	3,621
1881	595	1895	4,238
1882	599	1896	4,967
1883	844	1897	4,927
1884	920	1898	4,842
1885	1,080	1899	4,367
1886	1,607	1900	4,256
1887	1,865	1901	4,506
1888	2,212	1902	4,821
1889	2,413	1903	5,328
1890	2,505	1904	5,881
1891	3,137	1905	6,309
1892	3,282	1906	7,427
1893	3,661		

Source: Department of Commerce and Labor, *Report of the Commissioner of Corporations on the Tobacco Industry*, 3 vols. (GPO, 1909), 1:53.
Note: After 1900, the figure includes cigarettes intended for export.

The contextual causes of structural change become clear when we examine the responses that other firms made to the changing regulatory environment. In the late 1880s, for example, members of the Gravely family began suing one another over brand names. The Gravelys had been manufacturing tobacco since 1827, when they bought tobacco from their neighbors, a practice that continued into the postbellum age. About 1856, a relative had begun manufacturing tobacco nearby. In the antebellum business, that was no cause for alarm—the original manufacturers helped their kinsman gain a foothold for his enterprise. In 1886, however, when the sons of that upstart took up the trade, the heirs of Peyton and Willis Gravely sued to protect their firm's brands. Apparently the newcomers "simulated brands" of the established firm, using similar boxes, designs, and colors to package goods "of far inferior quality" to those they imitated. They "debased the reputation of [the original family's] tobaccos and trade-marks." Using brands to convey information directly to consumers, to fight off competition, to attach unique information to otherwise undifferentiated goods—more manufacturers than the Dukes were doing it.[48]

The defendants answered the suit by declaring that the leaf itself had the reputation, not the brands manufactured by any particular firm. Moreover, the brands manufactured by their father, the other Gravely, since 1856, had "contributed largely, if not chiefly, to establishing and maintaining the wide and favorable reputation of said Henry county tobaccos in general." The local agricultural product had begun to acquire its reputation, the sons thought, "at or about the close of the late civil war, say in 1864 or 1865." The distinction lay in the leaf "more than any manufacturer's mode of manipulating it," according to the defendants, and no manufacturer could defend a brand, such as Henry County Bright Pounds, whose identity relied on raw materials available to others. In fact, the regional produce had proved a resource for many manufacturers as they established and maintained the identity of their brands: D. J. Walsh & Co. stenciled their products as Golden Henry Co. Pounds, and R. J. Reynolds in 1876 made Star of Henry Bright Pounds. Qualities in the leaf, its bright color and the county of its origin, had begun to work as a generalized description of the manufactured products made from that agricultural commodity. Bright Henry County leaf served to distinguish those consumer products manufactured from it.[49]

This Gravely family lawsuit demonstrates the impact of the Civil War and the postbellum political economy on one family, whose economic interests had spanned agriculture, leaf dealing, and manufacturing since the 1820s. Merchants who packed together the crops of their neighbors, like Robert Wilson back in 1785, could no longer after the Civil War take their business to the next level by moving into manufacturing. Branding tobacco products and guaranteeing their consistency required dependable characteristics in their raw materials, which served as well to characterize the brands that manufacturers produced. Manufacturers in the postbellum period relied on raw materials to establish the uniformity of their branded, manufactured, consumer goods. While raw materials had characteristics that fell neatly into categories, especially place of origin and leaf color, technologies of cultivation and curing did not yet figure as economic information. Method of agricultural production did not yet fully distinguish one leaf from another. While taxing manufactured tobacco worked to differentiate agricultural from manufacturing processes, therefore assigning certain phases of production (curing and marketing especially) to the farmer alone, still identifying specific techniques with particular regions producing unique agricultural commodities would take more work. Achieving that would require farmer participation.

PART TWO

Constructing Varietals

A *yellow* leaf, *tough* and *silky*, is thought and said to be
perfection. Now can you tell me what chemicals will hasten
the growth and development of the tobacco plant; then what
will cause the plant to yellow on the hill?

W. J. C. to A. R. Ledoux, April 5, 1879

CHAPTER FOUR

Ripeness Is All

When the Gravely family fought and filed lawsuits for exclusive rights to brand-name manufactured tobacco goods, those brand names invoked agricultural products: Henry County Bright Pounds referred to where the leaf had grown and its color. Categories like these helped define the crop into varietal types. After all, postbellum manufacturers sought to create distinctive consumer goods, and consistency of finished products relied on consistency of raw materials. Yet, agricultural techniques changed after the war. In the fields of Reconstruction, particular methods performed in specific environments began regularly to produce recognizable leaf characteristics that, in the twentieth century, would become the basis of regulation. The technology reached closure and the taxonomy became set: a twentieth-century farmer who flue-cured his crop but found the results came out brown had not made something new, but had simply produced a bad crop of Bright Tobacco. The tobacco types evolved from changing institutions, just as the eighteenth-century Chesapeake system that made tobacco Virginian had issued from the inspection laws. The last chapter illustrated one change to the political economy of leaf production: taxation's impact on the tobacco industry and the consolidation of manufacturers into near monopoly. This chapter explores the solidification of cultivation methods into new systems within the new postbellum frameworks.

Taxation centralized the tobacco industry, but emancipation created multiple, concatenating contributions toward making tobacco bright. As slave plantations decentralized into small farms worked by individual families, one-time coastal commission merchants gave way to a new class of store-keepers. Local stores supplied farm workers with goods on time and gave landlords the mechanisms for paying laborers without cash. Providing wages to freed farm workers and supplying the crop's inputs required new systems of credit; paying off debts at the end of the year meant trying to finish production in December. In colonial times, it could take eighteen months, easily, for tobacco to move from the seedbed to sale. Industrialization accompanied independence and further fragmented the selling season into an all-year affair. This changed after the war. Cultivation, harvest, curing, and marketing procedures steadily transposed their timing toward the faster production of a finished agricultural commodity, to pay out the furnishing merchant and the hands, the landlord and the guano note. Fertilizer shortened the growing cycle and regularized production enough to trim the agricultural calendar down to very nearly an annual cycle. Harvest, curing, and marketing methods evolved to move the crop more quickly to its buyers, to close out credit provided on an annual basis.[1]

Emancipation and the shredding of merchant functions underlay the third and final institutional shift that brought the tobacco types into being: the rise of agricultural science, a new branch of the government. The informational work that middlemen had once done became tasks performed by new political institutions, some with regulatory power. Take, for example, the federal Department of Agriculture. The Morrill Act of 1862 founded the USDA during the Civil War. It began to cut teeth, however, only after about 1880. Before that year, its publications on tobacco cultivation bore the mark of antebellum assumptions. Early USDA articles rarely spoke of tobacco types; location authenticated the technological descriptions. That would change after 1880, when varietal distinctions became clearer and the USDA began to mediate markets between growers and buyers. Agricultural science as a government activity grew in the experiment stations established by the Hatch Act of 1887 and the extension programs founded by the Smith-Lever Act of 1914. The research that took place in experiment stations and in farmers' fields during cooperative extension contributed to the solidification of cultivation techniques that made predictable types. Despite its slow ripening, the USDA aided scientific enterprise that shaped cultivation systems.[2]

Taxation, emancipation, and the rise of agricultural science—these three influences intertwined, operating at one time. Together they constructed today's tobacco types. The remainder of this book attempts to untangle these changing institutions, to distinguish taxation's impact on the commodity chain from that of emancipation and the rearrangement of merchant functions. Making tobacco bright—defining varietal types and regulating markets through varietal designations—was a process that, like the inspection laws in the colonial period, wove together so many elements into one final result (Bright Tobacco, among other kinds) that the causes began to look natural. Credit arrangements and social structures such as landlords, tenants, share-croppers, wage laborers, and extended families joined with elite efforts to regain political power. Each shaped the emerging technological system of tobacco production. So did the transformations in the tobacco manufacturing industry, the consolidation of buyers, and the standardization of their needs. Technological change takes shape from contextual causes, and the changing post-bellum political economy wrought significant changes to cultivation systems, even on the smallest scale of individual farms and fields. Unpacking the causes and effects of changing tobacco technology is the task that lies before us.

EMANCIPATION

The emancipation scene deserves a higher status in American history—a stronger role in the national folklore, right up there with the Boston Tea Party, the surrender of the Confederacy at Appomattox Court House, the bombing of Pearl Harbor, the assassination of JFK, 9/11. Few slaves likely experienced the moment exactly as the legend tells it, but the image echoes through the ages. In the mythological version of the episode, the master stands on his white-columned porch and reads from a paper in his hand. In the historical imaginary, he is angry. He has always thought of the people gathered around him as his family, although their status differed from that of his sons and daughters. In point of fact, they have always been his slaves. He has called them servants, or children, or family, or by skin color—darkies or negroes or niggers—but never as slaves. That was a word for abolitionists, Yankee meddlers, British philosophers—people who never understood how hard his servants were to deal with, nor how seriously he took his own responsibilities. Finally, at this moment, he is correct: when the paper is read and the speech delivered, they are slaves no more.[3]

Legal freedom gave fresh opportunities to the workers whose sweat watered the entire commodity field. Emancipation initiated the most significant changes in the institutions and economics of Southern agriculture. Its impact, however, took time to work its way through the entire system, the whole commodity web created around tobacco. The same was true of cotton and corn: the sudden change in the legal and economic status of workers initiated slow-moving structural adjustments. In some regions and for some crops, the changing technological systems of agricultural production in the postbellum period meant that even the credit system took time to unfold: in November 1878, the *Southern Cultivator and Dixie Farmer* printed a sample of a sharecropper's contract when a South Carolina reader apparently wrote the editors to ask for one. For most Southerners, however, the contract arrangements that sustained postbellum agriculture emerged much earlier—for some, in the middle of the war, as Union armies became plantation overseers; for others, in the years immediately following emancipation, or military defeat.[4]

Masters and slaves both knew what freedom meant: the right to earn wages, the legal personhood that slaves had never possessed, the ability to sue and be sued, to own property and make contracts. What form those contracts would take, however, puzzled all participants. Sprigg Russell, for example, remembered quite well the start of May 1865, when his mistress (whose husband was dead) "called together all the negroes on the place, and informed them" they were free "and had right to make contracts." She "proposed to employ them for the balance of the year, to cultivate the crop then planted, and proposed to pay them in Kind [and] she would give them such a part or portion as the Regulations of the Freed Mans Bureau might require, or the Custom of the County adopt."[5] The promise to pay workers in kind, in part of the crop they produced, established sharecropping as a way for landlords who lacked ready cash to pay wages to workers. Sharecropping became the device by which workers extended credit to their bosses. By deferring their pay, they invested in the enterprise. Although sharecropping has a bad name, its principles were not cruelty: after all, according to the federal officials overseeing the region, that was the way the freedpeople wanted it.[6]

As Sprigg Russell's affidavit testifies, the annual contract between laborer and landlord that characterized both sharecropping and tenant agriculture appeared almost instantly upon emancipation. The turn of the year had always been a period of renewal, as masters had allowed some free time between Christmas and the New Year, and hired slaves had then returned home at the close of their contracts. The days before Christmas and the days after the New Year were likewise part of the holiday, making the fallow time in many cases stretch to some weeks. On some plantations, too, gift giving (and gift demanding) had cemented relations between planters and their bondsmen.[7] It made sense, therefore, that if contracts were to arrange the relationships between land-owning farmers and their emancipated workers, they should be annual contracts that began and ended around the New Year. The Freedmen's Bureau supported this arrangement, printing contracts and adjudicating disputes. Because landlords no longer owned their laborers, however, they no longer took any responsibility for feeding or clothing them. To get life's necessities, a population new to the free labor system had to rely on credit. Landlords soon learned, too, that credit for personal supplies could be extended to crop inputs, so that freedpeople could be made to assume some of the debt that agricultural production had always entailed.

The old blurry boundaries between the sectors helped some landlords assume the role of merchant to their workers. They supplied both agricultural inputs and the personal and household supplies that workers needed and counted such costs against the hands' eventual payday. After all, the *Southern Cultivator and Dixie Farmer* in 1876 reflected that "the cost of living inevitably regulates the price of labor" and advocated that "farmers" should supply "laborers" their provisions. These could either replace wages or lower their rates (if supplied "at *cash* prices") or give the landlord a storekeeper's profits.[8] Elias Carr was one who could follow this advice: born in 1839, married at twenty, he saw brief service in the Confederacy's military but spent most of the war at home on his East Carolina cotton plantation, under orders to provide agricultural goods to the war effort. After the war, he began to serve the function of merchant to his workers, keeping a store and providing both credit and goods to the new class of consumers created by emancipation of the slaves. As merchant-planter, he represented old ways and flexibility among the sectors. Yet, his turn to storekeeping and credit provision marked him equally as a New South type, a go-getter on the lookout for new entrepreneurial opportunities.[9]

Supplying goods meant acquiring them, however, and not all planters had the know-how or the credit to set up shop. The commission merchants who had supported production on a plantation-level scale since colonial times did not all see the wisdom in supplying smaller-scale producers who had no collateral beyond the crop under construction. When merchants ceased supply, or when planters took on the task, a new sort of firm would emerge to serve the newly emancipated economic actors. Tobacco had a particular facility for scaled-down farm operations. That should surprise no one who recalls that large-scale plantations first emerged as a result partly of the inspection laws and their need for labor, for the condensed tasks of harvest and curing. The postbellum crop returned to its seventeenth-century scale of production:

smallholdings worked now by tenant families, with stores to supply them and underwrite their efforts. There were also changes, however: few storekeepers helped market the crop as Robert Wilson had done for small farmers in 1785. Postbellum stores supplied small farm units but rarely sold their leaf. This old-fashioned yet new form of business demonstrates the impact of changing institutions on the political economy of each period of Southern agriculture.[10]

The new credit arrangements shaped the postbellum agricultural system. With landlords in debt to sharecroppers for their work and merchants for their inputs, with tenants owing rent to landlords and furnishing bills to merchants, with laborers of all classes in arrears to landlords and merchants for both production and personal necessities, the credit situation of the New South lacked both clarity and consistency in the first years after emancipation. Short and failed crops in the second half of the 1860s (and dropping prices) only exacerbated the difficulty. Credit risks increased as producers across the commodity web often failed to earn enough from harvest to pay the costs of production. The postbellum courts spent considerable effort establishing payment priorities. Legislatures soon stepped in to decide: when the crop was sold, who would first be paid? Merchants, landlords owed the year's rent, or farmers turned laborers who had deferred their pay until harvest? Unsurprisingly, workers fell rapidly to the bottom of this list. By the end of the 1860s, the legal foundation of New South agriculture—the credit system that agricultural production required—was solidified and ranked every other actor above the workers who waited for their pay.[11]

Postbellum merchants, landlords, and laborers established credit relations on an annual basis. As a result, because the crop had to be sold for anybody to realize its value, tobacco production began to shift into an annual production cycle, to fill a single calendar year. Freedpeople were too smart to do the tasks needed to start off next year's crop, a crop in which they had no economic interest. Making seedbeds—the task that began the production cycle, usually accomplished in the fall between harvesttime and the end-of-year holidays—was exactly the sort of work that they refused to do. The Freedmen's Bureau records are filled with such conflicts over the agricultural calendar, as freedpeople refused to do washing, mend fences, cut staves, or engage in any improvements to the farm that did not immediately translate into the production of a commodity crop.[12] The bureau's administrators understood that most freedpeople flat-out refused to work between the harvest and the end of the year. The sharp uptick in cases reported by federal administrators at the turn of every year, in December and January, demonstrated that conflicts over work were calendrical, centering around crop divisions and occurring at the close of contracts.[13]

The tobacco cultivation tasks that transformed most completely in the postbellum decades, therefore, were those related to the beginning and end of the production cycle, especially the cascading sets of tasks involved in harvesting, curing, and marketing the crop. Preparing a seedbed became a task most often performed in the New Year, although exceptions abounded for decades as the transformations worked through the system. Much of cultivation remained the same: from transplanting, to weeding and worming, to topping and suckering, many of the tasks established in the colonial Chesapeake persisted, although hilling slowly died away. Although it once counted as the limiting constraint on the scale of agricultural operations in the colonial Chesapeake, the practice slowly disappeared from postbellum tobacco fields. Harvest, curing, and marketing methods changed even more dramatically. These are the techniques that today define the varietal types: Bright Tobacco is flue-cured tobacco, which means that it has been cured with heat sent through flues (ducts) in the barn, protecting the leaves from smoke. It is also harvested by a method called "priming," which requires plucking the giant leaves from the plant as they ripen, one at a time, from the bottom up. These harvest and curing methods were mostly new in the postbellum agricultural system.[14]

Technological changes involve a multitude of individual choices. This is why historians of technology so often analyze whole systems of production to replace deterministic myths of invention. The harvest methods that took shape after emancipation fit better with emerging curing and marketing methods. When growers in the districts that became the Bright Tobacco regions adopted priming, stripping the commodity leaves from the plant as part of the harvest, the new technique ended the work of stripping that had once lasted throughout the spring. Priming did not really become a dominant harvest method until the twentieth century, however. One farmer, in the 1930s, recalled for the government interviewers that he "never started priming until about 1919. . . . We cured tobacco then on the stalk. . . . Raising tobacco then was different from the way we raise it now." The documentary record supports this late adoption of the method: there are pictures of "stalk harvesting" in 1897, for example. The new harvest technology of priming was recommended as early as 1870, however, for "those who cultivate tobacco on a small scale, or who have hands and time enough." The size of farms and labor forces shaped the technological choices farmers made, and ripe leaves became more important in determining harvesttime.[15]

The curing process also changed on the postbellum tobacco farms of Southside Virginia and North Carolina. As with harvesting, it took time for the technology to evolve. In fact, flue-curing had been available for decades when it finally began to standardize the leaf produced in the region and became the definition of Bright Tobacco. Folks had been patenting devices to improve curing methods since 1850 at least. Moreover, as the price bubble around Kentucky's Yellow Banks in the 1840s and 1850s had demonstrated, antebellum buyers had often wanted a mild leaf, more yellow than the colonial growths, to wrap plug or to provide a milder and more inhalable smoking tobacco. The curing devices patented in the years after the war demonstrated the desire, similar to that of the manufacturers, to profit from the peculiarities of growing particular commodities for specific segments of the market.[16] Nonetheless, the persistence of multiple curing methods, for any seeds grown, appears in an 1878 letter that one merchant wrote to a planter, hoping that "you sun cured and flue cured your crop this year," rather than taking other options. Postbellum buyers wanted leaf cured in specific ways: "You cant make any other sort of Tobacco that will buy you."[17]

AN EMERGING TECHNOLOGICAL SYSTEM

The development of these new techniques of harvest and curing satisfied the postbellum political economy. Harvest and curing methods fit into a technological system, analogous to the Chesapeake system in the eighteenth century. The tasks of Bright Tobacco production suited family labor: women tied (looped) the individual leaves onto the tobacco sticks, and men lifted the heavy-laden sticks into the barn for curing. A lengthy curing cycle involved adjusting the fires and modulating heat for nearly a week, which produced the fabled bright color in the leaf. Long curing fit into the periodic method of harvesting that plucked individual leaves from the bottom of the plant as they ripened. Priming and flue-curing together allowed extended families to rotate their labor across farms: "We barned Uncle Montgomery on Monday. Saturday was Uncle Dewey's day. All of us worked together. No one would barn Wednesday or Thursday . . . because that would mean adjusting the fire on Sundays and, well, this is the South." Individual leaves, plucked from the same parts of the plants, hung in one barn and cured together. Whole barns-full were therefore about the same size and color, aroma and order. So market-

Flue-curing barns were sealed up tight to keep the heat inside them, and fires were lit at the stone entries to the flues inside. The flues (once made of stone, these later became mass-produced ducts made of tin) carried the heat of the fires but protected leaves from smoke. This barn was photographed in Orange County, North Carolina, near Durham, in 1939, when the postbellum system for producing Bright Tobacco had solidified this particular curing method.
Source: Library of Congress, Digital ID 8c10578r.

———

ing began earlier: grading and tying into hands went more quickly with leaves of similar qualities. Stripping disappeared, too—or rather, took place during the harvest, replaced by priming.[18] To adopt one part of this system meant significant changes across the entire end of the production cycle, and several states stuck by older methods. Postbellum Maryland and Kentucky did not take up flue-curing. They maintained older harvest and marketing methods into the twentieth century. They still used hogsheads, as in colonial days, into the twentieth century.[19] Farmers who adopted the new technology needed new tools: Captain William Henry Snow sold his flue-equipped Modern Tobacco Barn as one part of new processes and viewed stalk curing as "a foolish waste of time and a wicked waste of wood." When farmer William Wallace White took up flue-curing in 1875 Warren County, North Carolina, stripping in the new year ended (except for the occasional crop of Oronoko he cured in older barns). Clearing out the new barns of flue-cured tobacco, to fill again and fire up, spurred his work and proved especially problematic in dry weather. Storing the cured leaf until marketing also required buildings, an enlarging element of the new system. The systematic structure of the changing technologies is demonstrated by advice tendered in a cultivation pamphlet: "Never cut before fully ripe, and enough fully and uniformly ripe to fill a barn." New harvest and curing methods fit together, solidifying both.[20]

Technological transitions in harvest and curing methods occurred slowly and unevenly, of course. The reorganization of marketing, however, emerged more easily and appears clearly in the new types of warehouse that sold tobacco in the region. Centuries earlier, the inspection laws had centralized tobacco sales at public warehouses and instituted the sale by sample of leaf in hogshead quantities. After the Civil War, tobacco warehouses became institutions at once more private and more public. Anyone with adequate credit and connections could open a warehouse, as the work of examining the tobacco and determining its value became the responsibility of the buyer, not a government inspector. Instead of the giant hogsheads, warehousemen arranged small piles of tobacco, still tied into hands but now displayed in baskets, for sale on the skylit warehouse floor. As in the rest of the agricultural system, the scale of operations became smaller. Auctioneers—colorful characters who worked for the warehouses, or traveled from tobacco town to tobacco town as independent agents—took the place of merchants in describing the goods. Buyers and farmers often met face to face as the chant was called, the price determined by bidding, the purchase financed overnight by the warehouse.[21]

Certain stories point to a shift in this direction before the war: Billy Yeargin has found examples of tobacco auctions from the 1820s and loose-leaf sales since 1858. Danville, Virginia, trumpets its antebellum experiences with auction sales.[22] Substantial evidence, however, demonstrates that the warehouse system of marketing seemed new to many observers after the war. For example, the R. G. Dun credit reporters, who had been wondering about tobacco's commodity web since the 1840s, were mystified by the marketing system embodied in the new warehouses. In 1878, the head credit-reporting office in New York City questioned the local correspondent's solidly favorable assessment of a Danville warehouse that had a three-thousand-dollar standing debt with a local bank. The reporter wrote back to explain the new structures surrounding leaf marketing: "The . . . warehousemen pay planters for the tobacco they sell as soon as sold by checks on their Bank. [T]hey

Marketing changes in the postbellum years meant that farmers sold leaf in small piles that buyers could examine directly, while auctioneers chanted the price until buyers finished bidding. For Bright Tobacco sold to domestic manufacturers, leaf no longer went to market in large hogsheads for colonial and state inspectors to break open, sample, and inspect. *Source*: Library of Congress image 8b28339r.

———————

pay therefore planters before they themselves collect of the persons who buy the tobacco[;] their Banks therefore advance the money by borrowing warehousemens checks. [H]ence the above agreement. [T]he Warehouse-men usually collect from purchasers on same day or day following & then the deposit makes good what the Bank has already on same day or day before paid over for the warehouse."[23]

The rise of the auction warehouse represents one obvious transformation of merchant functions in the postbellum political economy. Once, in the colonial and antebellum systems of tobacco cultivation and marketing, merchants had dealt directly with planters, the heads of enslaved or familial labor forces that operated as a single unit. Merchants provided planters with goods and credit on a large scale, rarely with a time limit, and sold their leaf a thousand pounds at a time, prized into hogsheads. When Robert Wilson packed together into one hogshead the crops of John Worsham and Captain Charles Williams in 1785, to send them to the breaks at Petersburgh, his work perfectly characterized the Chesapeake production system. After the war and after emancipation, sharecropping and tenant arrangements scaled down agricultural production into smaller units. Decentralized plantations required that storekeepers provide smaller-scale credit to tenants and sharecroppers, who sold their goods in those baskets on the warehouse floor. Private businessmen rather than public inspection officials, warehousemen matched up buyers and sellers and financed the transactions negotiated by the auctioneer within the warehouse. Warehouses also sometimes kept the information that underlay landlord-tenant arrangements for crop division and would pay both the seller and the landlord for their shares of the crop.[24]

Marketing fit well into the emerging technological system of Bright Tobacco production and suited the changing harvest and curing methods. A barn filled with individual leaves, harvested at the same time from the same place on many plants, cured in the same conditions, produced tobacco leaves of a similar size, shape, and color, which moved the crop more easily to market. The postbellum system of production harmonized in exactly the same way that the colonial Chesapeake system had. Each demonstrates the momentum of technological systems that economic historians call path dependency. Yet the elements in the technological system were not exclusively mechanical or methodological; they were organizational, legal, and cultural—credit arrangements and family labor structures included. Historians of technology call this the sociotechnical web: a seamless web of social structure, the technological basis for economic behavior. When engineers deliberately construct such a web, as when a civil engineer (for example) pushes to get laws passed that will enable her project to begin construction, the process has been called "heterogeneous engineering," because it encompasses so much more than solving physical problems. In the case of postbellum agriculture, no single engineer ever took charge. Still, the system became resistant to change as all the disparate methods cohered into one.[25]

While some planters (like Elias Carr) took up storekeeping in the postbellum South, others used the new storekeepers or warehouses to pay bills and supply goods to themselves and their workers, as planters had once utilized commission merchants. William Sutherlin, the Danville tobacco manufacturer, invested in farms after the war. He paid his workers by depositing half their monthly earnings at the store of A. B. Fowlkes, holding the remainder until harvest. The deposits paid for their purchases; those who actually received cash did so through the store. As tobacco warehousemen paid both landlords and laborers their portions of the leaf's price, so local storekeepers often likewise tracked the credit arrangements by which landlords and laborers produced agricultural commodities.[26] Such economic information gathered up in the towns of the New South, in the institutions that financed production as antebellum commission merchants had done. William Faulkner described a fictional version of this process in Frenchman's Bend, where Will Varner owned the local store, the cotton gin, the grist mill and blacksmith shop, and also most of the "small shiftless mortgaged farms" carved out of the wrecked plantation lands. Sometimes warehousemen and storekeepers were the same person, or the same family; kin or not, they embodied new versions of old-time merchant functions, and both usually participated in New South boosterism, town-building, and local government.[27]

Consider Jesse Willis Grainger (1845–1910), for example: Confederate veteran, farmer and landowner in Lenoir County, and the owner of Kinston, North Carolina's first tobacco warehouse, Grainger serves as an excellent model of the type. At "various times ... he was president of the Kinston Board of Trade, president of the Chamber of Commerce, vice president of two Kinston banks and one New Bern bank, president of the State Mutual Life Insurance Company, president of the Atlantic and North Carolina Railroad, and owner of the Kinston *Free Press*. Grainger was also the Lenoir County representative to the General Assembly in 1885, chairman of the Lenoir County Democratic Executive Committee, and a member of the State Democratic Executive Committee." Men such as Grainger were the founders of tobacco boards of trade that sprang up in towns throughout the region, as well as less-specialized chambers of commerce, local merchant banks, and various county finance bodies. Such organizations might host expositions, fairs that highlighted the crops and businesses of their regions. Expositions provided local elites with opportunities to work together and establish common bonds, to represent their region as possessing unique characteristics, special crops, and entrepreneurial towns, showing pride to the world and its markets.[28]

Men who opened up tobacco warehouses and joined or founded local commercial organizations made money by matching buyers and sellers. This meant furnishing buyers to farmers seeking markets for the commodities they produced. For this reason local boosters would write to the tobacco trust, the ATC, begging the combination to send buyers to their markets, their warehouses, their towns. "The quality of the Tobacco north of us is said to be very fine and a large crop: we shall need help to handle it and trust the American Tobacco Co will see it to their advantage to place a buyer here." Only a buyer could solidify the adoption of the crop, because without such an outlet, how could a farmer sell his produce? The successful cultivation of the commodity required the appearance of an interested procurer, and local elites across the Piedmont did not hesitate to ask the combination to send one. If such a request received a reply from the tobacco trust, the New South's tobacco promoters felt pleased: "allow me to thank you for your kind consideration of the matter of placing a buyer on this market," whether or not such consideration resulted in the desired appearance of the purchaser.[29]

CROP SPREAD

In opening up tobacco warehouses, promoting tobacco culture on their farms, among their tenants, or in the agricultural operations of their neighbors, New South boosters were responsible for spreading the crop into new locations, as Scottish merchants and their representatives had done two centuries earlier. Jesse Willis Grainger not only founded Kinston's first tobacco warehouse in 1885 but also served on boards of trade and chambers of commerce and ran local newspapers, railroads, and banks. He distributed tobacco seeds to nearby farmers as well, providing them the input to begin production. Moreover, his warehouse was the place to sell the article he had introduced and helped them adopt. Many folks who sought agricultural diversification—a classic cause of the New South leaders—found that tobacco served the purpose. For many farmers, cultivating tobacco meant employing the newest techniques for harvesting, curing, and marketing, the methods that made tobacco bright. This was when, for example (in December 1890), the Petersburg Chamber of Commerce submitted an advertisement in the *Progressive Farmer* newspaper calling for farmers to move from North Carolina into Virginia. "We have for sale cheap, a nice line of farming lands well adapted to the culture of Bright Flue Cured Tobacco," promised the ad. Spreading the crop meant spreading the technology, which helped establish the type.[30]

Bright Tobacco spread across the southeastern states in the decades after the Civil War and as the twentieth century turned. From a concentration along the Virginia–North Carolina border in the Piedmont center of the

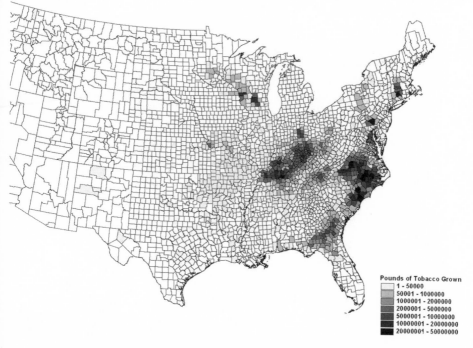

Pounds of Tobacco Grown
☐ 1 - 50000
50001 - 1000000
1000001 - 2000000
2000001 - 5000000
5000001 - 10000000
10000001 - 20000000
20000001 - 50000000

After the Civil War, Kentucky began to overtake Virginia as the center of both agricultural and industrial tobacco production.
Source: U.S. Census for 1880.

state in 1869 and 1879, the crop spread to the coast in 1899 and 1909. Nannie May Tilley's maps show production moving down over the border into South Carolina in just those decades. By 1919 and 1929 Georgia farmers were harvesting Bright Tobacco.[31] Tobacco cultivation was still spreading as the Great Depression descended on the nation. Many planters were like Elias Carr, who had turned storekeeper in postbellum East Carolina after emancipation and sought new crops to revitalize his worn-out cotton operations and to keep his laborers (and debtors) employed. In September 1890, he experimented with temperature variation in three flue-equipped barns of tobacco. Perhaps he was inspired by tobacco's appearance in nearby Wilson, as evidenced by a letter from a friend that a few weeks earlier had invited him to the "opening sales of tobacco at our new Tobacco Warehouse" in that town.[32] By 1898, boosters of East Carolina's tobacco capabilities declared that other regions "produce a bright tobacco, yet not the kind that enters into close competition with the bright cigarette tobacco of North Carolina."[33]

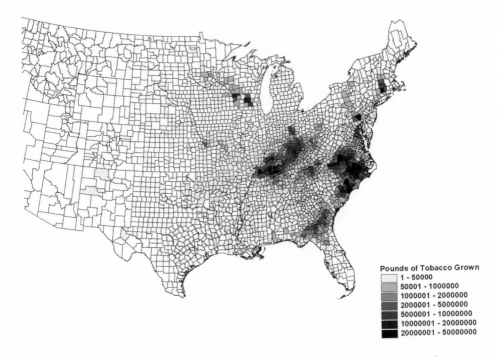

Pounds of Tobacco Grown
- 1 - 50000
- 50001 - 1000000
- 1000001 - 2000000
- 2000001 - 5000000
- 5000001 - 10000000
- 10000001 - 20000000
- 20000001 - 50000000

Tobacco agriculture spread into new regions as the tobacco industry grew and consolidated. At the same time, the regulation of tobacco production into particular types, defined by the technology used to produce them as well as the markets they served, meant that new types such as Bright Tobacco could be grown in more places. Comparing this map with the one on page 115 shows tobacco production expanding from Virginia across North Carolina, for example.
Source: U.S. Census for 1900.

As new regions adopted the technology and took up cultivating the crop, each stood to gain from counting its products as unique. Observers and buyers split types into the regions in which they were produced: Old Belt Bright Tobacco came from the origin point of the crop, on the Virginia–North Carolina Piedmont, where the hills began to rise to the west, toward the Appalachian Mountains. East Carolina was the New Belt and took up the crop only in the 1890s. Between them lay the Middle Belt, while production on the border of South Carolina was named the Border Belt. With the transfer of the technology into new locations came innovations: the size of New Belt barns was considerably larger than those of the Old Belt. Although each region made Bright Tobacco, the producers of each section benefitted when their products seemed uncommon. Noncompetition made distinctions among the

Bright Tobacco belts attractive, although "the classification of these is not based wholly on botanical points." The crop spread as boosters sought to diversify the economic basis of their regions. Even as typical methods moved from place to place, leaders in most localities claimed that their farmers made peerless products, trying to draw buyers to their fields.[34]

REGULARIZING THE METHODS

Fertilizer helped too. Not only did the application of fertilizer speed production and regulate plant processes, but efforts to sell fertilizer led to the spread of the method as an aid to producing the high-priced yellow tobacco. While guano had been available on the American market since the 1830s, fertilizing really only became common practice in Chesapeake and Piedmont agriculture after the war.[35] Emancipation sped its adoption in several ways, including sharing the costs and spreading the risks of new cultivation techniques across all the emerging classes of farmers, from landlords to sharecroppers. Many of the businessmen and boosters of the New South ventured into the fertilizer business, presenting historians with evidence of postbellum changes in business strategies and sources of wealth. The Thomas Branch Company was among them. It had been a classic antebellum Southern business, commission merchants of multiple partnerships, dealing in whatever came to hand—including tobacco manufacturing, in the various firms run by Peter McEnery. In the second half of the 1870s, the partners then running the firm dipped a toe into the newly profitable Southern business by forming the Cat Island Guano Company. It eventually abandoned the fertilizer business; its postbellum banking business was a purer form of commodity brokering and served the firm well for another century.[36] The use of fertilizer in making tobacco bright appears obvious if one compares Virginia with North Carolina, especially when placing the data of agricultural production next to that of fertilizer consumption (see table 4.1). In 1879, Virginia grew nearly three times as much leaf as North Carolina, on somewhat less than three times the land. Growers in each state, however, spent about the same amount of money on fertilizer, just over 2.1 million dollars—meaning that North Carolina farmers spent three times as much on fertilizer as Virginians did, per acre of tobacco land, and per pound of leaf produced. That crops other than tobacco must have absorbed some of the substance only strengthens the case. Crop rotation schemes that suited dark tobacco production left too much nitrogen in the soil for bright leaf. Sandy Piedmont Carolina soils starved the leaf of such nutrients and kept it

TABLE 4.1
Postbellum Fertilizer Use

	Tobacco produced (lb)	Tobacco acreage	Cost of fertilizers purchased ($)
North Carolina	26,980,213	57,208	2,111,707
Virginia	79,988,808	140,791	2,187,283

Source: U.S. Census for 1880, pp. 103, 246, 596.

light and bright while growing, but this made fertilizer necessary. Meanwhile, the methods that regularly produced Bright Tobacco made the application of purchased chemicals seem natural. Because Bright Leaf grew on sandy soils, it required fertilizer purchases. Fertilizer companies therefore pushed the flue-curing method relentlessly.[37]

Fertilizer companies were joined by seed sellers in spreading the technologies that made tobacco bright. Both sorts of firms used pamphlets as marketing devices, and these printed publicity materials linked advice on cultivation, harvest, curing, and marketing techniques with testimonials from growers whose authority apparently came from the region in which they grew: North Carolina farmers in Caswell or Granville County grew Bright Tobacco, and in the pamphlets they claimed that the seeds or fertilizer touted therein contributed to their success. These printed sales documents influenced the emerging belief that Bright Tobacco came from particular seeds, grew best with particular applications of chemicals, and required specific methods to produce the commodity. Some of the characters involved in spreading Bright Tobacco technology were not the sessile firms of narrow local interests. For example, Major Robert Lipscomb Ragland (1824–1893) tirelessly pushed Bright Tobacco production in the postbellum Piedmont. One of the principals in the Southern Fertilizer Company, which published for decades numerous pamphlets promoting the methods of tobacco production, curing, and sales that defined Bright Tobacco, Ragland also sold tobacco seeds that bore his name. Advertisements for his seeds, as well as arguments with the seed distribution service of the USDA, appeared occasionally in the *Progressive Farmer* newspaper that represented farmers' political activism in the late nineteenth century.[38]

Other stars in the constellation of New South business boosters also promoted methods of making tobacco bright, although those methods were not yet decided. Fertilizer helped carefully selected seeds grow on soils formerly considered poor and infertile, as the plaintive question to the experiment station ("What will cause the plant to yellow on the hill?") showed. Curing methods, however, would eventually become the dominant criterion for defining Bright Tobacco. Captain William Henry Snow, who sold both the Modern Tobacco Barn and the priming method of harvesting to prevent the waste of wood, was a friend of the Dukes, who headed the ATC trust.[39] Such associations powered the New South, as did the storekeepers and townsmen who founded warehouses, banks, and newspapers. Nonetheless, as late as 1898, even the curing method remained uncertain: one overseer in the Gravely family region, the heart of the Piedmont epicenter of Bright Tobacco production, wrote to his Hairston landlord, "you asked me which way it would yellow the best I think steam would be the best way to yellow it but I never tryed it but the Hotter and damper it is the better it yellows." The technology that made tobacco bright was still under construction, even as fertilizer companies, seed breeders, curing barn makers, warehousemen, and even storekeepers who stocked flues and twine for tying leaves onto tobacco sticks pushed the crop into new regions.[40]

BOARDS OF TRADE

Local boards of trade joined such men together and helped solidify the postbellum production system, as the Petersburg Chamber of Commerce advertised cheap land to Bright Tobacco farmers in the *Progressive Farmer* in 1890. Such organizations made the effort not only to bring the cultivators of profitable tobacco types into their region but also to find buyers for regional products, as well as to create and enforce agreements regarding fees and qualities and prices among the warehouses. One example can be found in "opening day," in which local warehouses, in order to accommodate the mobility of buyers and auctioneers, established dates that marked the beginning of the leaf marketing season for their locality. Indeed, "opening sales" such as that recommended to Elias Carr in 1890 became so common as the crop spread that they appear without remark in the historical record, even though the concept is hard to find in antebellum and colonial documents. Opening days of course also contributed to establishing an annual agricultural calendar. Such efforts by business leaders across the Southeast added to the momentum that would eventually solidify the notion that flue-cured tobacco was Bright Tobacco, a type with characteristics so distinctive as to function as a variety serving particular markets.[41]

This quasi-governmental regulation of trade demonstrates how constructed (to use a term from science and technology studies) were the seemingly natural components of Bright Tobacco, including the timing of tasks, the agricultural calendar, the use of particular seeds, and the impact of particular soils. Scholars usually employ the word "constructed" with the adverb "socially," so that elements of the experienced life—seen and heard, tasted and felt, measurable and quantifiable—can be shown as products of the social structure. Historians of course generally define the category "social" in terms of class, race, and sex, and these are indeed important elements of structure. They can leave out all-encompassing assumptions of culture, however. Nor do they explain those structural workings of the economy (supply and demand, the relative availability of land, labor, and capital, and therefore their value) that also construct the reality that humans experience. Yet economics and technology work in tandem. Technology sets the limits of the possible, the structures within which economic decisions are made. Poor sandy soils became more valuable wherever Bright Tobacco technology took hold. This demonstrates how factor endowments and economic fundamentals are contingent upon available technology. Technology too has causes: emancipation changed the agricultural calendar and, with it, tobacco's harvest and curing methods, making tobacco bright.[42]

THE ROOTS OF REGULATION

Quasi-governmental associations such as boards of trade and chambers of commerce exerted powerful pressure in mediating markets between buyers and sellers—but quasi-government is not the same thing as government. Regulations that rely on voluntary compliance and apply only to members are conventions, institutions without binding authority. Had agencies with some teeth not appeared in the federal government, Bright Tobacco might have been a passing fad in the tobacco trade. Still, although it did not yet have the power of law, the first appearance of varietals in federal discourse bears scrutiny. An early version of the taxonomy on which later regulation would rest appeared in the Tenth Census of the United States. The massive 1880 census classified the populations and productions of the nation as it emerged from its war of national integration and headed toward the second Industrial Revolution. Several individual volumes on agriculture and manufacturing each contained reports on tobacco, by professional investigators. The report on tobacco agriculture was based on growers' descriptions, as collected by a Tennessee statistician named Joseph B. Killebrew. The result—nearly three hundred pages of detailed, state-by-state reports of tobacco's agricultural practices and the markets for the products of each technological system—took the first steps that would later regulate tobacco markets by type definition.[43]

Killebrew's report for the 1880 census classified tobacco into types and set the stage for later regulation. Although he organized the text on a state-by-state basis and questioned growers about agricultural technology, his classification scheme depended mostly on markets. He gleaned his tobacco types from "Commercial circles," in which "classes" of tobacco were defined by "their adaptation to a certain purpose." These purposes mirrored old distinctions between shipping and manufacturing tobaccos: the three main classes of leaf consisted of smoking, chewing, and export tobaccos. Within those three classes, there were "types," which Killebrew based on "qualities or properties in the leaf, as color, strength, elasticity, body, flavor . . . or in the methods of curing." He then subdivided these distinctions into grades, "the different degrees of excellence in a type," the extent to which a particular product lived up to the traits of its type. Thus, the differences between tobacco types were entirely practical, dictated by markets: "yellow tobacco . . . is one type, yet it is used for both smoking and for chewing, and is therefore put into two classes; if exported, it would be put into three classes." Since the taxonomy depended only on demand, tobacco distinguished by one characteristic—its color—became several types occupying multiple classes.[44]

The census's reliance on market information should surprise no one. Of course, different regions prepared specific tobaccos for particular markets. Long-standing trade relationships had resulted in specialization between supply and demand, and Killebrew drew on existing categories of information that had developed in trade. Negotiations between buyers and sellers had long depended on similar categories to determine what the article was and what it was worth. In merchants' antebellum accounts, the location where leaf had grown represented half the information in transactions because, to buyers, where the leaf had grown indicated its other elements—its sensible traits, the look and feel of the leaf, the color and flavor and aroma, gumminess and chaffiness, and working characteristics. Of course, this might be a lie, as when English-grown tobacco sold as Spanish in the mid-seventeenth century.[45] Nonetheless, differences in prices indicated that buyers believed that the region of origin contained valuable information. Growers, on the other hand, knew where they lived. To them, information about crop qualities depended on its buyers—that was the other half of information in transactions. Classifying tobacco types according to market purposes established long-standing market relationships as if they were natural. By contributing an essay for each state, too, Killebrew strengthened the relationship between regions of origin and production methods in different areas.

Killebrew's state-by-state census essays captured technologies in transition. His information came from questionnaires sent to growers; in places, this meant that his descriptions rested on only twenty observations. Yet they represented actual agricultural and marketing practices, even if only those of a few growers. Some practices had persisted from the colonial period, but some had changed, or were changing as he wrote. In Virginia, for example, hilling seemed rare by this time. Transplanting bore the marks of emancipation since it took place almost always in January or later (although some planters, perhaps those who paid wages, managed to sow earlier). Transplantation, whether into "hills or drills," was still usually performed "after a rain," with considerable additional description of moisture requirements. Topping controlled the number and quality of leaves—yellow districts topped after the bud had appeared, for example, while dark-producing regions timed the practice by the number of leaves growing on the plant. Suckering received almost no coverage. The author assumed that it was performed as often as needed. Thus, the Chesapeake system of tobacco production lingered even as it changed around the edges, at the start and end of the production cycle. Hilling versus drilling (and thus the spatial control of labor) represented a part of the cultivation system still under construction, carrying profoundly local characteristics. It therefore received detailed treatment.[46]

Since harvest and curing techniques would later become the definition of Bright Tobacco, they deserve our close attention. Throughout Virginia, according to Killebrew's correspondents, harvesting tobacco still involved taking the entire plant, no matter the type. Growers "split the stalks and hung the plants astraddle of sticks," although some recalled that once upon a time, individual leaves had been harvested and cured. Killebrew wrote, too, as if ripeness were an objective, measurable reality—yet one that had only a tangential relationship with timing the harvest. "Tobacco cut fully ripe is not so bright, but makes a sweeter and better chew," whereas some found that "tobacco is more pliant and works smoother when not quite ripe," while in another county, "the forward plantings are cut when fully ripe; the later, when the leaves are nicely yellowed." Even in 1880, ripeness did not yet dictate harvesttime. Technical choice persisted, and methods of producing the "finest wrappers" differed from those "for the smoking tobacco," while heavy dews "benefit the shipping much more than the manufacturing types." Whether or not heat was applied during curing helped to define the type: in "the sun-curing belt," which grew smaller plants, "no artificial heat is employed."[47]

Sun-curing already referred to the product of a particular region, but most of the 1880 census still reflected a world in which sellers chose their curing methods—the era in which a merchant could still advise a planter how to cure his crop to make tobacco that would pay. Technological systems were still developing. "Some place the tobacco in the barn as soon as it is cut and hung, the two operations going on at the same time if the force is adequate." Yet older methods persisted, including scaffolding, bulking (which some thought did some of the yellowing), sweating, and wilting before curing. The curing barns of the various regions seemed to correlate with the specific types of tobacco they produced, but nonetheless, "the process of curing to be adopted by the planter depends upon the character of the tobacco as it comes to the knife." Flue-curing and charcoal-curing appear in the census as different "modes" of producing "bright yellow tobacco." Yet yellowing the leaf by curing was also part of the process of producing "red shipping," which "yellowed on the scaffold" before "firing," or curing by an open fire. Well-made charcoal apparently gave off little smoke when burned, making flues unnecessary.[48]

As for labor, it occupied a peculiar place in the census descriptions. In Virginia, for example, the labor system received distinct treatment on a county-by-county level, under the heading of the "present quality of Virginia tobacco." Leaf quality was analyzed in terms of past glories from which it had "deteriorated." Part of that was due to soil exhaustion, "agricultural depression," and "low prices." But part was the loss of skill and experience as "[t]he old 'hands,' trained in the operations of priming, topping, assorting, and the various details of cultivation and management," were "dying out." The "younger generation" Killebrew found "decidedly inferior to the old," although he did not specify that this was the cohort growing up after emancipation. Yet laborers influenced quality. "Tenants, the majority of whom are negroes, raise, as a rule, an inferior grade." Where "colored people, inexperienced and unskilled, who pay but little attention to the management of their tobacco," were the primary producers, the census found poor leaf. It could not explain who performed the actual work for those farmers who were "studying more carefully the needs of the tobacco-plant," nor those whose "spirited rivalry" competed for "who shall raise the finest tobacco and get the best prices." "Planters" made fine tobacco; freedpeople, only the poor grades.[49]

The census did not blame freedpeople for poor quality in North Carolina; there, "attempts to raise fine tobacco upon lands not suited to its production" could result in dark or shipping tobacco that was "not regarded as profitable." Type and grade distinctions mingled just as they had in eighteenth- and nineteenth-century price-circulars, not yet clearly distinguished from one another. Killebrew described bright tobacco as fine tobacco, a designation of quality more than of particular qualities such as color and flavor. He dated the "fine-tobacco interest" from about 1852 and attributed the yellow quality of the valuable leaf almost entirely to soil, "although care in the modes of cultivation and curing was also found to be necessary to the production of the best qualities." Before the war, according to his account, the "yellow-tobacco culture" first grew up "upon a ridge between two small tributaries of the Dan River," that Piedmont territory from which Robert Wilson had sent tobacco to market in 1785. Eli and Elisha Slade were the Caswell County planters responsible for spreading the crop "along the same ridge" into Southside Virginia. Although the census did not tell any myth about a sleeping slave who accidentally discovered a curing method, antebellum planters got the credit for spreading the method that made tobacco bright.[50]

As for seed, Killebrew noted that "slight differences in nomenclature, local names, and the uncertain use of descriptive adjectives make it difficult to reach absolute accuracy in treating of the varieties of tobacco cultivated." All the types that he listed "cure dark brown or red when grown on red-clay soils with heavy dark or brown top soil, but incline to brighter and lighter Hughes on sandy, gray soil, with yellowish subsoil, and cure from bright red to mahogany and fine yellow." Lands new to the culture made leaf grow brighter than did old fields. The association of the growing environment with the market purposes of the product seemed complete. In Killebrew's formulation, tobacco types bore no recognized relation to the seeds from which they grew, but soil played a crucial role. Changing land values reflected the soil requirements of good, light tobacco production, as "[l]ands worth from $1 to $3 per acre in 1860 now bring from $20 to $100," but, as is typical with cost calculations of new technologies, "nothing definite can yet be known, . . . the continued value of poor lands depending also on the stability of the demand for such tobacco." The value of inputs varies with the price their output can bring. Changing technology alters costs in unpredictable ways as the demand for its products rises or falls.[51]

Even as ripeness became more important, it still did not quite determine harvesttime. The plant could grow after it became "ripe," if wet weather forced a bit of extra growth. Curing plans dictated harvest more than any signals given by the plant, and harvesting still involved splitting the stalk and hanging it over a stick for curing. Harvesting occurred only "on Monday and Tuesday, so as to cure by Saturday, or it is cut on Friday or Saturday, postponing the curing till Monday, from the rigid regard for the Sabbath and its universal observance by all classes," although in poor weather this could cause "serious inconvenience and expense." Even more, although curing had become a method of making tobacco bright, more than one curing method could accomplish the deed. The Bright Tobacco of the census might be cured by open fires as well as flues, but the latter "promises very soon to supersede all other methods." Air curing had long been used to make the fine bright tobacco, but flue-curing allowed "better results, with more certainty." For Killebrew, curing served simply as a method of creating leaf that served particular markets. He understood its system-building properties: sorting while harvesting meant "that those only of a uniform color and ripeness shall be cut and cured together."[52]

Nonetheless, harvest and curing methods more and more distinguished different tobaccos. Louisiana leaf growers had long provided contrasts with the typical processes of Virginia and Carolina. Known to the census writer as Perique, grown only on two hillocks above the surrounding swamps of Saint James parish, tobacco technology in Louisiana had developed under a very different institutional framework than that of the Chesapeake. According to Killebrew's description, suckering was done until ripeness, and the plant was cut whole—no splitting of the stalk—and dangled over ropes hung in the shed.[53]

> Now begins the peculiar manipulation of the Perique tobacco. As soon as the leaves become embrowned, and while the stem or midrib is yet green, each one is carefully picked from the stalk and the green stem is pulled out. The first leaves are pulled off in about ten days from the time the tobacco is put in the shed, and from one to three leaves at intervals of a few days, until the whole stalk is stripped. As fast as the green stem is pulled out the leafy parts are made into loose twists, each twist containing from twenty to thirty half leaves. These twists are packed in boxes 11 inches square, capable of holding 50 pounds, which, when nearly full, are put under a simple lever press, the lever being 12 feet long, to which weights are attached, so as to secure a pressure on the tobacco of 7,000 pounds to the square foot. Screw-presses are never used, for the reason that a continuous pressure is required in curing this tobacco.

Growers first aired, then pressed again, and then cured again in this fashion for months, "until it shines in oily blackness" from "the juices" penetrating

A carotte of tobacco from the 1880 census, illustrating the unique way that Louisiana tobacco growers marketed their leaf. Their curing and marketing methods had developed in
eighteenth-century French Louisiana, outside the hogshead system that tobacco inspection laws had created in colonial Virginia.
Source: U.S. Census for 1880, p. 679.

the "whole mass." "Perique tobacco is cured and preserved by the resinous
gums contained in the natural leaf." It was marketed as it was cured, with
methods similar to those of the eighteenth century: it was made into carottes
on "winter and leisure days, and employs every member of the household."[54]

Like Chesapeake leaf, however, that of Louisiana was also defined by its
market purposes, its place of origin, and the techniques used to produce it.
The seeds of Perique Tobacco played little role in defining it: "Seed from Kentucky or from Tennessee makes a tobacco too rich and too large to cure well,
but if sown for several years in succession it gradually assumes the type of
that grown from the native seed." This unconcern with seed as varietal characterized Killebrew's whole essay. Some growers complained "that no care
is taken to preserve the purity of varieties," but his taxonomy cared not for
botanic distinctions. "The types and grades for which any given district or
section is especially noted depend not so much upon variety as upon peculiarities of soil, methods of cultivation, and subsequent management in the
curing processes." In certain districts, Killebrew noted that "generally much
care is taken to maintain the purity of the seeds of favorite sorts," but in
others, regular importation of new seeds made little difference in changing
the commodities the plants produced. The taxonomy of tobacco types that
Killebrew drew from his state-by-state analysis formalized categories that
had taken shape in trade, independent of their seed. Market purposes distinguished leaf from leaf; within those classes, niche markets, location of supply,
and cultivation techniques differentiated the commodities.[55]

The 1880 census printed Killebrew's taxonomy and sent a message: the
origin of the commodity correlated with other useful economic information.
To know one thing about tobacco, such as where it had been made, was to

know a whole host of other useful things, such as what had been done to it and what it was for. Moreover, the emerging definitions of varietal types most often relied on harvest and curing methods as the guarantor of type. As the agricultural editor of the Oxford (Granville County, North Carolina) *Torchlight* explained, "To raise fine Yellow Tobacco, first, grow your plants!" meaning, cure them later into the fine yellow types.[56] Connecticut cigar leaf, first produced in the 1850s, never riven by war and emancipation, was defined by its cultivation practices: it was shade grown. Bright Tobacco, a fine leaf beginning to make inroads in Piedmont, relied on cultivation methods first established in the colonial Chesapeake, although some techniques (such as hilling) established under the inspection system were slowly being weeded from the agricultural production system. Technologies of harvest and later processes would eventually define the type. In 1880, those techniques were still changing. In Louisiana, where few folks grew leaf but did so in well-established patterns, the ancient techniques persisted and made a recognizable and nameable product: Perique.

Historians of technology know that current technologies rarely represent the "one best way" to solve a problem or perform an action. Long experience and many case studies teach us that, most often, there was nothing inherently superior in the winning method that made it win, out of all the options available to solve the problem. The more closely that scholars examine the internal workings of the technology and its contextual causes as well as its effects—the harder historians look at the moment we call technological closure—the less it appears that the physical, natural world explains the solution. Instead, contests and controversies mark technological choices, as different groups express diverging interests and disparate solutions to the problem under consideration. The priming method of harvesting Bright Tobacco worked not just for physical reasons—a ripeness that crept up the stalk, but had never before played a role in cultivation decisions and technical strategies—but rather because priming served multiple functions for laborers and landlords, as well as buyers and manufacturers. It fit into an emerging technological system that, combined with specific cultivation, curing, and marketing requirements, reliably made the yellow tobacco that so many interested parties called "Bright."

How this technological system for producing Bright Tobacco reached closure was no abstraction. It incorporated all the elements of postbellum American history. Reconstruction ended in 1877, but the timing of Redemption (when the one-time planter class regained political control) varied from place to place. In 1880, when J. B. Killebrew provided the first tobacco taxonomy from the federal government, neither the organization of labor nor technologies of production had quite yet settled into a recognizable, predictable system. The vagueness of some of the census descriptions can be partly explained as Killebrew filling in gaps between the observations his respondents supplied—historians surely know the feeling. Part of the reason for his uncertainty, however, was because the political economy had not yet become the solid structure it would take at the turn of the century, when it came to seem natural, as political power coalesced in ways that prevented laborers from challenging their bosses. This was a story not just of the South, and not just of tobacco production: challenges would arise in cotton and wheat fields, too, and in Northern industry as well. Our version of the story, however, and the coalescing of political and economic power into its twentieth-century forms, takes shape in the fields of tobacco agriculture and the curing barns that made tobacco bright.

CHAPTER FIVE

Inventing Tradition

Each tobacco type embodies a response to the shifting political economy. The changing postbellum industrial structure—its narrowing from an intricate web to a simple commodity chain, with federal laws eventually limiting who could sell to whom—ultimately permitted the regulations that brought the tobacco types into being and defined them by the technologies used to produce them. The process was slow and uneven. In North Carolina, the Farmers' Alliance found a political foothold in type definitions and technological barriers to entry; Bright Tobacco served as an organizing framework by which rural leaders challenged the political parties and rising corporate power. The People's Party failed in its national electoral efforts, broken on the back of the two-party system and its sectional divisions. Populists did, however, achieve government support for agriculture in the form of new economic and scientific institutions. Moreover, farmers' radicalism did not end with the Populists, but extended into the twentieth century. It took time and work for the new industrial structure to develop, and agrarian unrest characterized the process. Tobacco farmers responded to consolidation among their buyers with political activism, and sometimes violence. They branded their crops, using government power to establish varietal types that they thought would increase their market power.

The Perique tobacco of Louisiana represents one response to the changing industrial structure of the late nineteenth century. As taxation separated agricultural tasks from industrial processes, the techniques of cultivation, curing, and marketing that had developed in eighteenth-century French colonies—curing the leaf under pressure, and twisting it up to sell in carottes rather than hogsheads—so closely resembled manufacturing that the federal government taxed Perique as manufactured tobacco. Its growers, naturally, protested. "It is not a chewing tobacco," Louisianans fumed to the federal government in 1882; it "is too strong to be used by itself." Nowadays the USDA calls it a type. It enters certain pipe tobacco blends, and the American Spirit company offers a cigarette brand that uses its exotic flavor.[1] Its markets are so narrow and production so small that Perique production and marketing have retained certain characteristics of its colonial political economy. In 1971, it still took four or five years to move from seedbed to final sale. The merchant-farmer relationship remained unusually close: merchants paid the growing and processing costs for farmers, supervised the final stages of curing and handling, and controlled the quantity farmers produced and the prices they received. The factor's close relationship with the manufacturer, also persisting for generations, guaranteed the market demand on which farmers relied.[2]

Perique is but a single example. Tobacco farmers in other regions also responded to the changing industrial structure and the combined power of leaf buyers by accepting a characterization of varietal type on which they could rely when selling their goods. Although perhaps unaware that their actions branded their commodities, the farmers helped to develop those institutions that defined the tobacco types. Type designations ultimately rested on three categories: region of origin, market purposes, and techniques of production—harvest, curing, and marketing methods in most cases. Farmers used political activism to fashion their identities and their traditional rural cultures around the stability of their commodity crops, and their actions built schools and scientific societies that allowed these newly established traditions to achieve regulatory force. A large body of literature in cultural studies has provided the title for this chapter. Case studies in country music, for example, demonstrate how expressing authenticity mattered when selling songs. Fashion is another subject within which scholars have seen expressions of modernity in the use of older forms, sometimes borrowed or imagined. Rural worlds are notoriously susceptible to this process, and the imagined golden age of the past seemed more vital to farmers who felt pressed into an uncertain future.[3]

This chapter picks up two threads from the last two chapters—the postbellum transitions in both industry and agriculture—to weave them back together. It describes a process, to borrow the jargon from the scholars of technology, in which relevant social groups in the tobacco trade constituted themselves into stakeholders. As the tobacco manufacturers entered into a near monopoly, farmers recognized the unfortunate mess in which they found themselves and responded with political activism until finally their governments took action. Tobacco's most important stakeholders were growers and buyers. Growers now operated on a smaller scale, whether one counts as "growers" the landlords or the farm-working families. Buyers had changed too, as the commodity web contracted from myriad firms doing modular tasks into the few companies consolidated under the influence of Civil War–era taxation. Merchants had once mediated the relations between the two. This changed in the postbellum era, as storekeepers provided credit, warehouses matched buyers and sellers, and both kept the information about landlord-labor relations that underlay agricultural production. This new system of financing the crop and marketing its products fit with the smaller scale of postbellum agricultural production. However, other forces, as well as individuals' actions, also helped bring the tobacco types into being.[4]

CHANGING INDUSTRIAL STRUCTURE

Turn-of-the-century tobacco farmers often blamed the ATC for the low prices their crops brought. Their anger at their buyers represented a common farmer complaint in that era. Business historians have generally seen this period of industrial consolidation in terms of firms employing strategies of horizontal and vertical integration: the first, buying up or destroying competitors doing similar business; the second, controlling distribution, or raw material production, or both. Yet individual firms did not necessarily employ integration strategies in deliberate ways—and the descriptions are false not just for tobacco. Andrew Carnegie allegedly achieved his legendary control of steel production because (according to most scholars) he seized control of ore as well as turning it into steel. In fact, however, Carnegie himself had no interest in ore. He fought his firm's acquisition of ore fields and facilities. Only a subsidiary's interests—and the eventual force of his board—drove Carnegie into the actions later described as a deliberate goal. Likewise, the ATC rarely grew tobacco leaf in the United States, but still its impact on tobacco farmers' markets ran deep. Our revised view of the combination, which sees its origins in taxation rather than the mechanization of production, provides a fresh perspective on farmers' response to its formation.[5]

In fact, many of the ATC's most effective tactics mirrored the decentralizing tendencies in tobacco's agricultural sector, although its general trend was combination and therefore consolidation. Advertising, as already remarked, put brand preferences for the first time in the hands of consumers, which contributed to the combination's control of distribution. Brands of manufactured tobacco had before the war spoken principally to merchants, signifying the durability of the article as well as its flavor, color, and aroma. The brands launched or acquired by the ATC, however, became known to consumers, even demanded by consumers, some seeking to complete those sets of illustrated cards. Joe Camel and the Marlboro Man grew directly from these efforts to create a loyal following among users. Of course, consumer demand for particular brands meant that the ATC could tell not only distributors and wholesalers but also retail storekeepers what brands to carry. As a result, the ATC transformed the merchandising of its goods. Rural merchants had lost much of their old market power while the tasks they once performed now took place in new institutions—the country stores, cotton gins, and tobacco warehouses of the postbellum rural South. Likewise, the ATC created specific needs among storekeepers and thus diminished the power of wholesalers, higher up the chain than individual shops, to determine what stock ended up on whose shelves.

Brands gave the Dukes and their partners dominion not only over wholesalers and retail shops but also over the raw materials they bought to make those products. The ATC found that the segmentation of consumer goods into brands worked to break up leaf into types. The ATC was itself a combination of several consolidations of the producers of specific tobacco products: the American Snuff Company, American Stogie, American Cigar, and the Continental Tobacco Company for plug tobacco. Such "sharp distinctions in character between some of the minor branches of the business" motivated the "separate corporate existence of minor tobacco manufacturing concerns." Then, each firm in the ATC had its own products, its own brands, and its own share of the markets for those products. Every ATC tobacco product, whether snuff or cigarette, came labeled by a brand—and each brand required dependable raw materials to maintain its own consistent characteristics. "To enable us to make our order to Mr. Kearney more explicit in the future, both for 'Duke' strips and 'Preferred Stock' cigarettes, we would thank you to write us the grading of same, whether by number, or letter." Categorizing the types and grades of tobacco used in the various brands would be crucial for establishing a purchasing strategy for the combination as a whole.[6]

It proved wildly difficult. The firm's efforts to select and set prices on its raw materials ran into difficulties more typical of the antebellum industry.

The qualities of the leaf and the various criteria for its price resisted standardization even after the purchasers had consolidated into a trust. The ATC certainly benefitted from the postbellum transformation of the agricultural sector, however. As emancipation and annual credit provided a set calendar for leaf to reach market, a series of opening days across the producing regions, it became possible to schedule buying, to minimize salaried workers without much chance of missing out on good leaf. As cultivation, harvest, curing, and marketing techniques became associated with specific regions and the market towns that served distinctive agricultural hinterlands, the ATC could rationalize buying by associating certain types with particular locations. Nonetheless, the process, although structural, was hardly automatic. It took work. Even the ATC combination, buying raw materials in all the major markets of the United States, found that decisions about price and desirable characteristics referred to the brands for which leaf was bought, and thus to consumer preferences, more than to any abstract quality which all the buyers and manufacturers recognized and understood.

Benjamin Newton Duke, the stay-at-home brother of Buck Duke (who ran the whole ATC operation from New York City), was responsible for the "Leaf Department," as the purchasing branch of the ATC was known. Ben Duke sketched out the Leaf Department's structure in the spring of 1891, to begin purchasing through the summer and fall, as the new leaf began to reach the market. The first organization of the Leaf Department dedicated an employee (some with staff) to each major market and also arranged for buyers in the smaller towns, covering nineteen locations altogether. Some of these, like Danville (Robert Wilson's neighborhood in 1785), were marketing centers for unique regional produce—in this case, the Henry County tobacco familiar from the Gravely family's Bright Pounds. One general manager, J. B. Cobb, oversaw the entire purchasing operation. He traveled by train from one tobacco town to another, examining the available leaf, consulting the local ATC buyers, and studying the firm's purchases. He spoke of his initial efforts with the combination's buyers in the individual warehouse towns: "I can only start them on the general grades and add the others as the markets supply them." If particular desirable characteristics had not yet appeared in the crop for sale, its qualities were too enigmatic for him to instruct his staff.[7]

Grading was always tricky and subjective, even for the tobacco industry. One fellow, offering freelance services to the firm, stated his abilities to judge leaf and the potential difficulties in doing so for others: "I have had six years experience in handling tobacco & I think I know the difference in first & second class cutters. . . . If you find my judgement on tobacco to not coroberate with yours you to have the right to cancil the contract." But the firm needed distinctions more specific than those between first- and second-class cutters, as each constituent firm had its own brands, and each required specific characteristics in its raw material. It became necessary, finally, to "decrease the long list of leaf marks, by condensing" the grades that brands required. "[G]et a sample of each mark, forward same to the Home Office or some central point, and then call before the Leaf Committee the practical leaf man from [each of the] Branches and see if some of the marks can be consolidated." Buck Duke himself suggested that "a sample of each mark, together with the cost price on Ware House floor, be held at [the New York] office as a guide in determining the approximate cost of the leaf used in the respective products of the Company."[8]

The grades of leaf that each branch of the business bought referred to the brands it produced, its sensible qualities such as color, and the way it would work through the manufacturing process, rather than any reference to how the leaf had been grown or cured, or its varietal heritage. Centralizing tobacco grades and types, across brands and across the markets in which it was bought, helped the ATC set leaf prices.[9] Location contributed to the expected characteristics, as the Leaf Department's organization demonstrated. Marketing towns, where local middlemen had built warehouses for tobacco sales, served as descriptions of types into the twentieth century, and the needs and organizational strategies of buyers help explain why. It took most of the first season in which the combination bought leaf as a unit to make decisions about how leaf characteristics could be standardized for purchasing purposes. But when the task was done, it was done: "we did the needful as regards buying tobacco on the different markets and making the several grades and think all is satisfactorily arranged. . . . The samples which have been sent to New York will be the types wanted and no doubt will answer the purposes asked for."[10]

Lucky Strike, according to the 1940s advertisement, "means fine tobacco." But this brand was neither the first nor the only one to advertise its raw materials. Another product of the R. J. Reynolds firm was the "Apple Sun Cured," a chewing tobacco defined by the curing method of its raw material. Sun-curing a chewing tobacco associated old-fashioned curing with old-time ingestion, which together likely bolstered the wholesome image of the brand. Another Reynolds product, the Camel cigarettes introduced in 1913, famously blended domestic with Turkish tobaccos. The brands of the ATC itself sometimes used the raw material as a means of identifying desirable characteristics: the Half

& Half brand sold a mixture classed as half burley and half bright. Many products of the late nineteenth century blended the Western Leaf (with its superior flavor-absorption capabilities) with the Bright Leaf (of yellow appearance
and rich flavor). Imported Turkish tobaccos contributed to the exotic identity
of cigarettes before they came to dominate consumption around World War
I. Tobacco types served well the new consolidating version of the tobacco industry, which needed to rely on particular agricultural commodities to make
consistent branded products. Types and grades also helped the firm control
its outlays for leaf, while still leaving the risks of raw material production in
the hands of farmers.[11]

Despite all the care taken in the purchase of raw materials, the Dukes
never involved the company in substantial agricultural operations—at least
not in the United States. In 1901, the end of the Victorian era, the ATC invaded the British market, sparking a rapid consolidation among manufacturers there, "a defensive amalgamation." This firm was named the Imperial
Tobacco Company, and it fought briefly against the ATC's standard policy of
buying or destroying its rivals but a year later saw valor in surrender. In 1902,
the Imperial and the American Tobacco Companies united into the British-
American Tobacco Company and split the world between them—employing
uniform grades and types for leaf purchasing around the world. In far-flung
locales, where this multinational corporation sought to produce cigarettes for
local markets and to supply them with local agriculture, the firm did promote
technology transfer and the agricultural production of specific tobacco types.
The British settlers in Rhodesia and Nyasaland, for example, who took up
tobacco cultivation and flue-curing in Africa, received aid from their fellow
merchants and imperial scientific institutions. As America likewise expanded
from continental acquisitions that became states in the Union to its new island protectorates, the ATC found that producing leaf locally contributed to
successful cigarette manufacturing. By 1906, for example, it acquired a controlling interest in the Porto Rican Leaf Tobacco Company.[12]

At home, however, the ATC made little effort to grow or cure its own leaf.
It had no need to integrate backward into raw material production. Controlling the characteristics, quality, and price of its crucial agricultural input was
unnecessary when so many small producers shouldered the risk of getting it
wrong—and enough farmers appeared to sell their goods in enough outlets,
all the emerging tobacco belts, the tobacco towns of postbellum Carolina,
Virginia, and Kentucky, that the Dukes were well supplied and spoiled for
choice. The different growing regions began to acquire distinctive character-

istics: technologies changed a bit in new locations, as New Belt barns of East Carolina were often built bigger than those in the Old Belt of the Piedmont. Opening days for individual markets meant that buyers judged the value of leaf from particular locations relative to that market's produce. Widening markets evoked product differentiation, as local boards of trade touted regional characteristics, and merchants selling flues, fertilizer, and other agricultural inputs joined landlords and laborers who wanted remunerative work in the soil. All these people and organizations sought to establish the desirable tobacco types in their regions. They became stakeholders in defining the traits of Bright Tobacco.

Thus, the formation of the ATC changed the markets for tobacco farm products. Taking shape in the 1890s as a result of federal taxation that separated agricultural from manufacturing processes, segmenting the consumer market into brands and leaf into types and grades, providing consumers with the power to decide what brand to buy, and thereby controlling distribution and demanding that wholesalers and shopkeepers carry their goods, the ATC represents a structural transformation of the tobacco industry (see table 5.1). While scholars know this period of tobacco manufacturing history quite well, its combinatory power, its global alliances with like-minded British firms, and its eventual dissolution as a result of the federal antitrust case together mark these decades as unique. As a centralized buyer that made purchases based on its own products and distinguished tobacco types by region and town, the combination altered tobacco agriculture. Community leaders eager to find markets for local products pushed farmers into adopting methods that reliably made tobacco of particular color and characteristics. Local merchants transformed their local economies by selling flues, giving away seeds, and building warehouses. As we have seen, they welcomed the ATC, writing letters to its leaders and trying to arrange purchases in their towns. These efforts, as well as the increasing reach of the ATC as the leaf buyer, swayed tobacco farmers to meet their markets in ever more specific ways, making limited products with distinctive characteristics sought by manufacturers for particular brands.

FARMER RESPONSE

None of this was lost on the farmers, whose reaction to the rising consolidation among their buyers took well-known, historically recognized forms. From the Grange through the Farmers' Alliance to the People's Party, which

TABLE 5.1
The Twentieth-Century Tobacco Industry

	1900			1910		1920			1930		
	Cigars	Mfd	%	Cigars	%	Cigars	Mfd	%	Cigars	Mfd	%
California	1,888	0	0.7		0.8	10,767	74	1.1	2,115		0.2
Connecticut	1,776	0	0.7		0.7	4,305		0.4	41,087		3.3
Florida	10,891	0	3.9		5.2	37,926		3.7	3,274		0.3
Illinois	8,741	3,168	4.2		5.3	11,827		1.2	8,653		0.7
Indiana	2,537	58	0.9		1.0	10,066	279	1.0	25,918	179	2.1
Kentucky	1,507	14,948	7.7		4.5	3,581	20,548	2.4	3,768		0.3
Louisiana	1,407	1,084	0.9			8,221		0.8	1,776		0.1
Maryland	2,843	7,054	3.5		2.5	8,190		0.8	6,074		0.5
Massachusetts	5,298	0	1.9		1.9	11,560	331	1.2	15,103		1.2
Michigan	5,589	3,746	3.3		3.9	18,119	9,574	2.7	1,381		0.1
Missouri	2,746	25,101	9.8		7.4	2,830	44,922	4.7			
New Jersey	2,648	7,788	3.7		5.8	32,299	24,419	5.6	75,585		6.1
New York	49,028	4,632	19.3		18.4	163,105	2,688	16.4	81,669		6.6
North Carolina	230	13,621	5.2		8.6	226,636	33,188	25.7	480,039	1,076	38.5
Ohio	11,240	5,753	7.4		6.9	28,854		2.9	16,549		1.3
Pennsylvania	31,483	1,247	11.8		12.0	98,371	5,423	10.3	108,061	4,409	9.0
Tennessee	291	1,541	1.1			590		0.1		16,872	1.4
Virginia	4,844	10,708	7.5		6.1	63,273	15,077	7.7	147,702	17,745	13.3
West Virginia	1,060	1,363	0.9			4,075		0.4			
Wisconsin	3,256	1,632	1.7		1.5	5,889	4,119	1.0	1,844	126	0.2
All other states	12,328	1,393	4.0		8.8	23,178	78,629	10.0	46,295	138,926	14.9
Product totals	161,631[a]	104,837[a,b]	99.9		101.3	773,662	239,271[a]	100.0	1,066,910	179,333	100.0
U.S. total		266,468[a]					1,012,933			1,046,242	

Source: U.S. Census for 1900, 1910, 1920, and 1930.

Note: The values of Cigars, Mfd, and U.S. totals are in uncorrected thousands of dollars. "Mfd" stands for Chewing, Smoking, and Manufactured Tobacco and Snuff; after 1880, the separate category for "stemming" or "stemmed" tobacco has been added to these. In the twentieth century, the census counted cigarettes and cheroots among the "Cigars" category, although the internal revenue kept more detailed figures.

[a] This total has been calculated and does not agree with the total presented in the census. "%" has been calculated as a percentage of this calculated total, rather than the one presented in the census.

[b] The data for "Mfd" tobacco include the figures for "Tobacco, Stemming."

sputtered in the presidential politics of the 1890s, the lineaments of agrarian radicalism have institutions and moments familiar from textbook histories. Founded in 1867, the National Grange (of the Order of Patrons of Husbandry) has generally been accounted for as the precursor organization and assigned in the cabinet of memory to the more decorous Midwestern wheat-growing regions. The Farmers' Alliance, founded in 1876 Texas before taking off like tumbleweed across the southern landscape back into the Southeast, often appears as the product of cotton monoculture and its reply to industrialization, railroads, and threats to an established way of life. It is folded by history into the next stage, the People's Party, the third-party political challenge to the Democratic and Republican machines. The traditional story of farmer activism reaches its climax with the election of 1896 and the defeat of William Jennings Bryan as Democratic candidate for president. Historians categorize these farmers as either traditional or ambitious. Were they looking back, trying to hold on to ancestral ways in the face of wrenching change, or were they forward-facing maximizers seeking the best deal out of a new economic order?[13]

Neither: these farmers were inventing tradition in the service of change, and their adoption of postbellum tobacco cultivation and curing methods and government-designated varietal types helped them along. This was the period in which the story of the sleeping slave first appeared: the Farmers' Alliance newspaper, the *Progressive Farmer*, reported Stephen's appearance on a warehouse floor, where someone recognized him, "pulled off his hat and gave a cheer," which the whole crowd took up, "and the old man grinned all over his face for five minutes." This was 1886, and the purpose of the myth, as recounted in the newspaper, seems plain: "Stephen belonged to Elisha Slade of Caswell, N.C. He has always voted the Democratic ticket, and says of his old master, 'I wish he was alive to-day and I was his slave.'" The postbellum decades saw many myths like this one come to light. An idyllic plantation, full of happy slaves, seemed for many white Southerners to lie just behind the ghostly screen of the war, beckoning with symbols of nostalgia for the past. At the same time, however, both regional pride and industrial competition lured farmers into the future.[14]

The tobacco types all have myths to justify and explain a technological system into normality, even Perique: one twentieth-century USDA expert denied the influence of colonial French settlers on Louisiana's typical leaf-curing techniques, so different from the Chesapeake system. Many other growers had been known to twist up tobacco that had been soaked in its own juices and cured under successive applications of pressure. White burley, too, had emerged from Civil War–era Kentucky complete with its own tall tale. The mutant leaf "colored well and produced the pounds . . . but was adjudged bitter to the taste." Yet markets shaped its acceptance: when "shipped to the Cincinnati market [it] sold at a high price."[15] Nowadays, of course, it is harder to draw such distinctions. Burley is different from Bright Tobacco, according to the most recent gene sequences, but not by much. There are genetic differences, to be sure, among the types. But "manufacturing quality traits" such as "yield, plant type, leaf quality, and disease resistances" appear to match up only imperfectly with genetic distinctions. Aroma, too—of such interest to those who buy leaf for the products it will make—has from the early twentieth century been recognized as the product of enzymatic action in fermentation, not genetics. Processing matters at least as much as breeding in the production of commodities from crops.[16]

Myths of varietals and their origins served farmers better than mundane reality when they relied on the unique qualities of their crops to organize themselves into political alliances. Farmers of particular commodities identified themselves as interest groups and voting blocs and pressed for government aid. Tobacco farmers led the way. The Farmers' Alliance first reached sunlight in tobacco country, not in the cotton districts that sowed its seeds. The weekly Alliance newspaper, the *Progressive Farmer*, was published in North Carolina. Its editor, Leonidas Lafayette Polk, was one of those New South characters of colorful name and useful disposition. He piloted the Farmers' Alliance from protest into politics and led a movement of radical class resentment, headed by farmers, into a contest for the presidency. Polk's death in 1892, just as his organization began to breach party walls, marked the end of third-party hopes. Polk and his followers tried to speak to all producers, at the bottom of every commodity chain. In the end, they organized their movement into suballiances based on unique commodities and sectors instead of uniting all farmers and workers along the lines of common economic interests. Such divisions contributed to the ruin of their plans.[17]

The insurgency of tobacco growers diverged somewhat from the ordinary run of farmer politics. For one thing, it is not very hard for today's scholar to sympathize with their anxiety. Cotton and wheat farmers feared middlemen now distant from modern experience: cotton gin operators, the sellers of the jute bags in which cotton was usually marketed, the operators of Chicago's huge grain elevators, the railroad corporations that transported both crops to distant markets. In the case of tobacco, however, we may more easily conclude that the consolidation of the manufacturers into one giant organization

Air-curing tobacco on the stalk in Kentucky, 1916. This curing technique completed a cultivation system that differed dramatically from the technologies of Bright Tobacco cultivation in Virginia and North Carolina—where, in the twentieth century, individual leaves were harvested for flue-curing (see p. 109). Differences in harvest and curing methods had begun to define specific types of tobacco, since they resulted in such dissimilar products, serving distinct purposes, and entering unique markets.
Source: Library of Congress, Digital ID nclc 00528.

created too much market power for the buyers, and that the centralized decision making of the ATC's Leaf Department did damage competition for farmers' leaf, the rivalry that might at one time have kept leaf prices high. Market segmentation for manufactured tobacco products did seem to diminish competition for the raw materials that composed those goods. Different types and brands of consumer products sought distinct characteristics in the leaf. Each brand wanted a different type of leaf, so buyers were not quite competing with one another. Except for the very best grades and types—bright leaf included—the Leaf Department had divided its purchases in such a way that it could set prices and farmers either complied or could not sell.[18]

Tobacco growers responded to the consolidation of their buyers by building their own brands. Ordinarily, the pioneers of branding were the large-scale producers of consumer goods: Nabisco, Quaker Oats, Ivory Soap, as well as ATC. Farmers hardly fit the mold. Whether analyzed at the level of tenant or landowner, farmers ran none of the "massive advertising campaigns" for consumers typical of brand creation. Yet, if those advertising campaigns were sometimes attempts to create barriers to entry into the market by smaller firms with fewer resources, farmers made branding efforts analogous to those of corporations. They participated in identity politics that relied on economic interests and identified themselves as Bright Tobacco farmers, or cotton or wheat farmers. They used political activism—and eventually, some of them, violence or anticapitalist radicalism—to express their dissatisfaction and to try to increase their market power. Other political institutions—especially the federal Department of Agriculture (the USDA)—helped them achieve brands for their products. So did their buyers, from the Gravely family to the Dukes. Manufacturers used raw materials to identify their goods to consumers, building the distinctive identity of their own products and also those of their raw materials. These emerging agricultural brand identities employed familiar categories to convey information: place of origin and market purposes.[19]

Technology helped brand the leaf. The lead article of the first issue of the Farmers' Alliance newspaper, the *Progressive Farmer*, initiated a series explaining how to "manage" tobacco. It reflected the transformations in cultivation wrought during Reconstruction: "from the plant-bed to the warehouse," the subtitle assumed that the product would be sold in small piles typical of the postbellum private auction warehouses. Advice for the plant bed and the timing of its preparation already incorporated the agricultural calendar that had developed around the credit arrangements of emancipation: "though the time extends" from mid-November to the start of April, "we prefer to burn from the 4th of January to the 1st of March." The difference between New Year's Day and four days later seems obviously a man-made divide, even if the difference in weather between November and January appears more natural. As always, however, the technology of agricultural production took at least as much direction from its social and economic context as from the requirements of the plant. Part of that environment was the organization of production, the structure materializing at the turn of the century—including workers' expectations of winter holidays, free from work.[20]

The *Progressive Farmer* articles consulted experts, "successful growers and handlers," and explained their techniques for producing "fine bright tobacco." No one mentioned seeds as specific to the type. Location of origin had long verified the characteristics of the crop. The first USDA descriptions of tobacco production had not specified the types of tobacco being produced—the concept did not yet exist.[21] The sources of the newspaper articles were becoming more commercial, however, and the commodity they produced more plainly named. The Durham Fertilizer Company had solicited the cultivation tips printed in the paper from a Granville County, North Carolina, family "who were selected to write the treatment . . . because of their skill and success in raising fine tobacco and who are endorsed by all of the Warehousemen of Durham as 'among the best and most successful growers and handlers of fine bright tobacco in the State.'" The first article appeared in February 1886. Installments followed throughout that year, with instructions timed to appear at the right moment, to meet the needs of farmers producing within the postbellum calendar of tobacco agriculture. The newspaper sold the method. Captain Snow, too, used the results obtained in the state agricultural experiment station to advertise his Modern Tobacco Barn in the *Progressive Farmer*.[22]

The Farmers' Alliance also used the tobacco type as part of its organizing strategy. The coordination of farmers according to agricultural commodities for the purposes of political action appeared, for example, in the announcement for an Alliance meeting in January 1890. The convention advertised was "mainly in the interest of the tobacco growing counties," and Alliance gatherings from "all tobacco counties" should each choose three delegates to send. At that convention, delegates gathered from the "Farmers Alliance and Industrial Union of the Bright Tobacco Belt of North Carolina and Virginia" became a "Special Permanent Organization," one created specifically to "defeat the purpose of any combination that operates to depreciate the price of leaf." The Bright Tobacco type was becoming recognizable outside the circles of local organizers. It was in the pages of the *Progressive Farmer* that the Petersburg Chamber of Commerce advertised its plea for farmers who knew how to produce Bright Tobacco to come into Virginia, transfer the technology, and grow the desirable leaf in a new region. Religion helped organize farmers, as did the commodities they produced, and other local power structures contributed as well. Commodities, however, were the elements of the organizational structure with which the groups were named. This contributed to the stability of tobacco varieties and the establishment of distinctive methods in particular regions meeting specific markets.[23]

Such small groups and organizational efforts were the vertebrae of an emerging spine of agrarian political activity, one that reached for national electoral office in the 1890s but failed. While the history of the Farmers' Alliance has been somewhat too closely coupled to the fortunes of the People's Party, the outlines of that story correctly sketch the one association as precursor to the other. The Populists' efforts to act powerfully on the national stage did not succeed, and the People's Party endorsement of the Democratic candidate for president in 1896 helps explain the reasons why. William Jennings Bryan was a powerful orator for those issues agrarian radicals held most dear. His "Cross of Gold" speech remains required reading for an audience that has forgotten how the gold standard obsessed the Gilded Age analysts of federal monetary policy. By pegging U.S. currency to the limited supply of gold held in national coffers, the federal government limited the cash in circulation. This kept commodity prices low. Farmers wanted the government to use silver too, in similar ways, and thereby increase the currency at work in the economy. Some even wanted greenbacks untethered to precious metals, a currency whose value could vary. They turned to third parties to advocate these goals.[24]

The failure of the Populists lay in the sectionalism that characterized American politics after the Civil War. Party games depict the methods each party used to retain its power in a particular locality for specific elections. Local power players used gerrymandering, bribery, dividing time, poll taxes, and literacy tests to discount votes and disfranchise inimical voters, keeping a solid South in the Democratic camp for nearly a century. At the same time, the selective admission of new western states helped northern Republicans preserve their Electoral College majority in presidential campaigns. Paying attention to intricate and local-level detail while analyzing the national political scene can illuminate the way structures stood on individual agency and regional culture. After all, the political parties differed in the North and the South. In the two sections only a generation removed from the bloodbath, the political parties whose solidification had helped bring about the Civil War meant different things to their constituencies. In the North, the Democrats seemed representative of the little guy, the factory worker and farmer and small producer who struggled to set more currency loose in the economy, to finance those hard steps up the ladder to self-sufficiency. In the South, however, the Democrats were the planters, a party of entrenched power in land and access to both labor and capital.[25]

The Republicans likewise had different constituencies in the North and the South. Although neither party was willing to abandon the gold standard, the northern Republicans were more thoroughly goldbugs and the party, as always, represented large entrenched corporate power. Even Southerners who had been Whigs before the war remained unlikely to join the Republican Party: the Black Republicans, the party of Lincoln, the party of Negro suffrage and African-American threats to established political power. Reconstruction branded the Republicans an occupying army, the bureaucracy attempting to control the economy by establishing free labor systems of production. These sectional divisions within the Democratic and Republican parties doomed the third-party option the Populists constructed. Their attempt to merge with the existing parties was foiled by those parties' dual nature. The founding fathers were right that political parties were unstable coalitions of interests, and in this case the coalitions held true because in each region different blocs were represented. The tragedy of the farmers was their inability to unite with manufacturing workers, as historians have long claimed—but this divide was sectional as much as sectoral. In the South, Democrats were planters and Republicans represented freedpeople. In the North, Democrats were factory workers and Republicans represented corporations. No third party could straddle those gaps.

THE SUCCESSES OF AGRARIAN RADICALISM

Radical farmers found success in places other than the presidency, however. Their community-based movements contributed not only to that national political activism that fizzled so noisily in 1896; they also catalyzed new institutions, which took over many of the functions once performed by merchants. These institutions provided significant government support to the agricultural sector, to farmers and their way of life. These props included, but were not limited to, rural free delivery from the national post office; the nation's agricultural marketing and credit systems (albeit devised with input from both growers and buyers); the federal Department of Agriculture (the USDA), newly empowered after about 1880, with its later programs of cooperative extension; and the federal system of state land-grant universities devised to serve the public interest. These Big Government supports for a single sector continue today. The social safety net that farmers enjoy does not extend to the cities, however. Price supports for commodities and national funding for scientific research into agricultural problems—these institutions of federal power help agriculture alone. For the tobacco farmers of the Chesapeake and those who shared their technology, Big Government established tobacco types as the foundation on which price floors and market-smoothing legislation could rest.

The history of North Carolina State University typifies the new institutions of the era and the role of those organizations in the emerging definitions of tobacco types. As the tobacco culture shifted into North Carolina and more growers adopted fertilizer as an input required by their technological choices, the state began to address the vulnerability of farmers to the blandishments of the chemical companies. Extravagant claims enticed farmers into buying useless products, so in 1877, the legislature of North Carolina hired a chemist to assay fertilizers to test their sellers' claims. From this modest beginning sprang a host of institutions that represented farmers' interests and, along the way, helped define tobacco types in the late nineteenth century. The State Department of Agriculture, founded in 1877, was the brainchild of the same Leonidas Lafayette Polk who later established the North Carolina Farmers' Alliance, edited the *Progressive Farmer* in the 1880s, and died before the Populists could nominate him their candidate for president. In the 1870s, he was merely one more one-time slave owner on the small-scale North Carolina model, a veteran of the Army of Northern Virginia during the Civil War, and alumnus of the Confederate State Legislature. He became the first commissioner of the State Department of Agriculture.[26]

Polk's multiple roles as farmer advocate and community organizer gave him the position from which he joined with New South boosters to establish those government institutions that made tobacco bright. In the 1880s, Polk gathered "a series of mass meetings of farmers to demand the establishment of an agricultural school." At the same time, the Watauga Club—a group of young men from the state capital of Raleigh, "dedicated to progress," as they described themselves—contributed their support to found a school devoted to agricultural and industrial education. Federal funds had been set aside for such schools under the Morrill Act of 1862, which, like taxation, had begun to affect Southern states as they stumbled home into the Union. North Carolina undertook to form such a school for the first time in 1887, although the bill made its way through the government only in 1899, and the board met for the first time in that year.[27] What became North Carolina State University came out of the attempts of the state to warrant fertilizer for farm use and to help farmers better meet their markets. Regularizing cultivation techniques was one way to do this.[28]

Government-sponsored agricultural science—in state agricultural societies, A&M (or "land-grant") schools, and of course the USDA too—took over many of the informational tasks once performed by merchants. If the commercial sector's physical and economic work of matching buyers and sellers and providing credit for transactions had become lodged in the private to-bacco auction warehouses of the New South piedmont towns, the informa-tional work merchants had once performed also needed scaling down to meet the demands of a different technological system of production. No longer was a letter from a merchant with a careful description of the goods a reasonable justification for the price it had brought. Now more abstract categories, to be applied more broadly, were the means of defining market demand and value, the criteria for quality. Moreover, these new institutions existed specifically to help farmers meet their markets and generate more profits; agricultural science standardized production methods specifically to achieve those goals and aided farmers in growing just what manufacturers demanded. Yet these institutions did not emerge on their own, from the changing structural de-mands of the transformed political economy. Individuals helped, and local se-lection explains why the results were so different in the different regions.[29]

Leonidas Polk was not the only figure to emerge from the Farmers' Alli-ance to carve new institutions into the landscape of the postbellum tobacco economy. Elias Carr did too. The East Carolina cotton planter turned mer-chant after the war, supplying goods and credit to his tenants, expressing older blurry boundaries among the sectors. In 1890, he became the man responsible for the spread of Bright Tobacco culture into his district, the coastal plains of eastern North Carolina, the worn-out cotton lands whose owners sought crop diversification. At the same time, he was riding farmer unrest toward political power, running for North Carolina governor in 1892, becoming the Democrat's candidate who successfully defeated the Republi-can-Populist fusion ticket for state leadership. Carr's leadership of the state Alliance, which dated from 1889, surely established his credentials as one who spoke for the farmers. So did his position as trustee of the land-grant college that later became North Carolina State University. His correspondence as Alliance president demonstrates the persistent efforts of the organization to limit membership to farmers, striking from the rolls anyone who opened a store, for example, or owned a tobacco warehouse.[30]

Elias Carr and his correspondence demonstrate the importance of farmer identity in political organization and in the historical record. Defining some people as farmers and others as not was an action Alliance officers regularly performed. It is also common practice among historians. The historical record tends to call the landlords farmers and to portray the families who worked the land as laborers, while modern scholars prefer to call the workers farmers. Recognizing class distinctions between those who owned the land and those who worked for them brings the development of the postbellum cultivation systems into clearer focus. Inventing tradition involved establishing tobacco cultivation methods as well as new institutions. A large part of changing a technological system involves managing labor, whether self, family, hired, or enslaved. With emancipation, the new methods that had emerged once again rewarded small production units more typical of tobacco production before the eighteenth-century inspection laws, smallholdings now rented out to operatives for shares or cash. Such farmers participated in a new technological structure for tobacco production that served the changing postbellum political economy of credit, tenancy, and auction markets.

BLACK PATCH WARS

In Kentucky and Tennessee, violence attended the emergence of tobacco types. Tobacco farmers traced an arc from activism to savagery. Most scholars ascribe the Black Patch Wars, as the conflict became known, to a dark fire-cured tobacco unique to the border region, to the isolated and traditional society its cultivation methods fostered, and to the clash between traditional culture and consolidating buyers.[31] This interpretation seems inadequate to the events. Farmers began by pooling their crops, to use the local terminology for cooperative marketing. They formed organizations and pledged to pool their tobacco together, to withhold the crop from the market long enough to drive up the price. The most successful of these, the Planters' Protective Association (PPA), worked well—as long as local banks supported the effort and lent money to farmers who stored their leaf in the Association warehouses. Not everybody pooled his leaf, however. Some farmers kept their crops outside the cooperative and then benefited when the shortage drove the buyers to offer higher prices for all available leaf. Conflicts over pooling revealed the deep-seated divisions in rural society, which then escalated into violence. Disparate classes of farmers had different economic interests, and the brutality of the Black Patch Wars stemmed from these divides.[32]

The PPA was founded in September 1904, as the leaf markets opened and farmers found that the offered price was half what they hoped. The next year, night riders assembled by the PPA—"riders on horseback, silhouetted by burning barns"—blew up tobacco barns and warehouses in small Kentucky towns and harassed the factors who represented distant leaf buyers. When federal tight money policies weakened local bankers' support for the movement, the actions of the masked horsemen turned malicious. In December

In Kentucky in 1906, most farmers still sold their tobacco in hogsheads. As curing techniques characterized the unique products of different regions, so did Kentucky's hogshead markets contrast with the warehouses of Bright Tobacco sales, in which buyers examined small piles of tobacco displayed in baskets on the warehouse floor (see p. 111). These disparate marketing methods had historical causes but looked natural in the way they shaped cultivation methods, work arrangements, and thus economic fundamentals (land values and labor costs, as well as the provision of capital for production).
Source: Library of Congress, Digital ID 4a13353r.

1906, 250 night riders seized the town of Princeton, Kentucky. Hooded and armed, they rode into town, cut telephone and telegraph contact with the outside world, overwhelmed the police force and the fire department, and dynamited one tobacco warehouse and set fire to another. Half a million pounds of raw leaf burned furiously as citizens cowered in their homes. Although tobacco fetched higher prices the next year, the frenzy continued. Night riders burned Hopkinsville, Kentucky, in December 1907. The timing was no coincidence: the goal was to burn tobacco bought from farmers outside the pool, but the fire spread to the Association warehouses holding tobacco off the market. Night riders did not shrink from beating and sometimes murdering resistors, witnesses, and sometimes just vulnerable black laborers peripheral to the conflict.[33]

Historical treatments of these events usually begin with the formation of the ATC trust. The "urban-based national and international business organizations" were "extending their reach into the countryside." The low prices they offered, according to most interpretations, threatened a traditional way of life producing a distinctive type of leaf. As a result of this predation, the "price a farmer could command for his produce declined for no apparent reason other than the hunger of huge 'trusts' for bigger profits." This seems too simple an explanation. After all, urban-based firms had sold country commodities to distant markets since colonial times, and agricultural producers rarely get to set the prices of their goods. Nonetheless, growers had good reason for dissatisfaction. The price of Kentucky leaf had dropped in the postbellum decades and plummeted after the cigarette manufacturers combined into the ATC in 1890. Kentucky leaf brought nine cents a pound in 1866, right after the war had ended, and eight cents a pound in 1888. Ten years later the price was half of that, and four cents a pound paid the farmer less than he had spent in producing the crop. Across the region, farmers sought to organize against the trust by pooling their crops, refusing to sell until shortages forced buyers to pay more.[34]

Historians who blame the monopoly echo the opinions that growers left in the historical record. Contemporaries recognized that the Italian *regie* (the government monopoly that dated from at least the nineteenth century) and the American Snuff Company—one of the branches of the ATC—had in the twentieth century well-established noncompetition agreements that kept leaf prices low. The postbellum period and the tobacco combination had in fact reduced the number of buyers at work in the region. In a damning and oft-quoted anecdote, a buyer testified that according to the rules of his employer, he was not able to cross the road to bid on a barn of leaf that fell outside his allotted territory. The ATC and the British-American Tobacco Company had split world markets among their constituent firms. Reduced competition did mean fewer buyers for the region's products. Growers had understandable motivations, and historians' interpretations reflect their reality.[35]

The segmented market of the ATC, however, made leaf prices more complicated a story. The brand-specific needs of the industry meant that it sought unique characteristics in the leaf it bought, and in the decade before 1909, no region of Kentucky made one single kind of leaf. The central part of the state was known for white burley—the new type that had appeared during the war, with its own myth of discovery and its own traits that suited particular purposes. Burley had become an input in Kentucky's particular kind of plug, that made by the ATC. R. J. Reynolds had also begun to use it in smoking tobaccos (first Prince Albert, a pipe tobacco sold in cans; later, in Camel cigarettes). The ATC used burley in some of its smoking tobaccos as well, including its Half & Half brand. The new white burley that had appeared during the Civil War proved a useful component of cigarettes as they began to take over the consumer market in the twentieth century, for the way it absorbed flavoring. Burley had a wide range of uses. Other Kentucky tobaccos entered narrower markets: the dark leaf still went to Africa or Italy or into the mixtures snuff makers wanted. Preparing local leaf for specific distant markets made it unsuitable for other buyers. Agricultural specialization contributed to the limited competition for the crop.[36]

Kentucky tobacco had its own essay in the 1880 census, and this revealed a remarkable range of leaf types produced in all parts of the state. The writer of the census essays, J. B. Killebrew, divided Kentucky into eight individual tobacco-growing districts. Five of these districts later participated in Black Patch Wars. Each district grew many, many kinds of tobacco—not just the dark fire-cured leaf associated with the conflict. Paducah, for example, also known as the Western District, grew tobacco classed into five separate market categories in the census and more than ten individual types. Some were dark shipping tobaccos, but others were bright wrappers. A portion of dark fire-cured tobacco still sold in Africa. Some European governments also bought dark Western leaf and controlled its distribution in their nations, the Italian *regie* among them. A large number of leaf types produced in the Black Patch region did each enter specific markets with limited competition. No region, however, could be described as having a specific traditional culture associated with production of a certain type of leaf. The range of tobacco produced in each district was too wide and various for that to be true.[37]

Moreover, tobacco types did not yet firmly or formally exist in the 1900s and 1910s, when Kentucky's tobacco wars raged. Farmers' depictions of their products, as their manuscript returns to the census indicated in 1909, remained variable and indeterminate into the twentieth century. Asked what type of tobacco they grew (along with their status as owner or tenant, their labor relations, and what kind of fertilizer they applied), growers referred to their crops by old nicknames. Fred F. Williamson of Henderson County, Kentucky, for example, grew "Hang down or Mar[y]land twist back" varieties, while one share tenant of that county who did not sign his return clarified that "most all tobacco grown in the stemming district is of a dark type fired & unfired. It is claimed that the unfired won't stand the trip across the sea." James H. Fulton grew "Anderson Prior. and allso Yellow Mamoth. cured in open Air." Another unsigned return stated that "mine is dark fired," but that others in the county grew leaf he called "Bright." Historians describe Henderson as a dark fire-cured leaf-producing district during the Black Patch Wars. In 1909, however, numerous varieties and curing methods reigned.[38]

This point deserves emphasis: farmers' tobacco types differed from the government's, and from historians' characterizations. In western Kentucky's Trigg County, Alex Wallace grew "Lizzard Tail" tobacco, while J. W. Adams was "reliably informed" that Trigg County made a "Dark Heavy export Tobacco" whose "best grades [were] taken by the Italian Government." As for his own crop, however, "I plant a variety known in this country as Little Yellow." Among the returns from Clay County, Missouri, where "Night Rider Depredations drove [planters] out" from Kentucky, Chas. Tho. Fisher said, "Tobacco here is put in 6 different Grades: 1. Plyings 2. Trashes 3. Luggs 4. Brights 5. Dark Red, and Tips___(the more Bright, the better the price all round)." Back in North Carolina, in Caswell County, the birthplace of Bright Tobacco (according to the traditional myths), Rufus B. Dabbs produced cigar wrappers and fillers in addition to flue-curing Bright Tobacco. He included in his census return a slip of paper declaring, "Other Tipes ar grow Which I am not Well prepared to give information on at this time." Other writers seemed more certain: "We grow orinoco tobacco Warren Hop good yellow prayer in Caswell Alamance Orange Durham & Granville Counties." Even this correspondent, however, said the type he grew was called Warren tobacco (named for a North Carolina county), but he classified it as Maryland air-cured.[39]

This was 1909: very late in our story for the seed or varietal type chosen by the farmer to be so insecurely associated with the curing technique, the place he grew his crop, and the market in which he aimed to sell it. Kentucky tobacco types, however, had always been extremely various. Back in 1858 and 1860, the Kentucky Agricultural Society gave awards for best quality in the "shipping" or "manufacturing" categories. Similar competitions took place in Kentucky warehouses after the war. One postbellum warehouseman took pains (as a merchant might once have done) to explain to a client that his hogshead of "Blk. Wrapper" could not compete for the prizes offered: southern Kentucky versions of that description differed from those that came out of West Virginia. One dealer "regretted that there was not a distinct class for West Va. alone" in such competitions. "There should be a Separate Class for her or that Southern Ky. should be ruled out entirely." These growers still packaged their goods in hogsheads. Different qualities of the same type came from specific regions, and warehouses classified them differently for competitive purposes. One can presume that buyers also understood these distinctions and planned their purchases accordingly.[40]

Commitment to tradition in tobacco production techniques cannot explain the Black Patch Wars. The 1880 census described a wide range of leaf types produced in that district—not just dark leaf. It was also the region from which David Bullock Harris and Alexander Barret had exported the Yellow Banks in the 1840s and 1850s. Unique kinds of tobacco and the production methods required to produce them can explain very little. Nor can the sudden intrusion of international buyers into agricultural commodities markets account for the violence. Selling tobacco downriver to New Orleans, where it was inspected for sale to world markets, had typified tobacco sales near the Kentucky-Tennessee border for nearly a century. Certainly the Black Patch region had a tradition of violence supported by the religious commitment to powerful paternal authority, as one recent history has drawn the culture there. Many backcountry places share these characteristics, however, without erupting into war. Evangelical religion also motivated many North Carolina farmers to join the Alliance and provided organizational structures for their political action. Yet North Carolina's tobacco towns were not seized by armed horsemen and burned for warehousing leaf.[41]

New questions might evoke different answers: What distinguished night riders from their victims? What differences among the region's tobacco growers led some to kill others, scrape their plant beds, burn their barns, and dynamite the warehouses? Some folks pooled their tobacco; others kept their crops out of the cooperative and benefited from the shortage it caused—theirs was called "hillbilly" tobacco. Hard feelings against growers who employed this strategy fanned the flames of ill will. A farmer who sold his crop outside the cooperative in the fall of 1905 might be forced to scrape his plant bed the next spring. Landowner and tenant/sharecropper status also divided Black Patch farmers from one another. Kentucky farms worked by tenants rather than owner-operators increased between 1880 and 1910 from a quarter of the total number to more than a third. The number of tenant farms, however, nearly doubled. Some farmers had mortgages, too, and the dropping prices of their products threatened their positions as the independent yeoman farmers that historians portray. Cooperative marketing also kept landlords and tenants bound together longer, awaiting the sale of the crop from which both expected payment. Tenants who wanted to move on from last year's landlord to a fresh contract had incentives to sell, rather than pool, their crops.[42]

Race marked another distinction among Black Patch tobacco growers. Robert Penn Warren, poet, novelist, Southern agrarian, and Pulitzer Prize winner, was born in the town where the Planters' Protective Association got its start. His *Night Rider* novel raises the Black Patch Wars to the level of literature. In that book, the hero received threats warning him to evict African American tenants in favor of white families. When he failed to comply, his house was burnt down. In real life, in March 1908, homegrown terrorists chased the entire African American population from Birmingham, Kentucky. The Louisville *Courier-Journal* reported the incident and concluded that "tobacco stemmeries have been warned to discharge negro hands and have complied and many landlords have turned them out. The same policy has been pursued in Lyon County" as the racial violence spread. Raids across the region drove African Americans from their homes. Political power was at stake: "redeemed" to Democratic rule as early as 1867, Kentuckians prevented "negro domination" of the polls with lynchings and assassinations. Political power depended on class as well as racial disfranchisement. As C. Vann Woodward knew, behind white supremacy "there raged a struggle between Southern white men" of different classes, with divergent interests.[43]

White elites needed labor to work the land. The president of the first cooperative association in the region said in 1904, "In our country we have an ignorant class of laborers who know nothing except what they learn from us. We white people teach them all they know and take care of them." He also described the dark tobacco cultivation system as "the most slavish crop" he knew, requiring constant effort. Indeed, in 1922 the USDA proved him right. It found that labor requirements for Bright Tobacco spiked at harvest, curing, and marketing. For burley, the work swelled at stripping and marketing moments. The tasks of producing dark fire-cured tobacco in Kentucky, however, were more evenly distributed across the year, a typical technological choice of slave owners, who had sunk costs in labor and therefore had incentives to make work even in slack seasons. The 1880 census essay echoed these associations.[44] Immigration advocates, too, who dreamed of attracting the populations then entering Northern factories, had interests in denigrating available laborers. A conference pushing immigration had been held in nearby Nashville, Tennessee, just three weeks before the night riders invaded Princeton, Kentucky.[45] Keeping labor in line through violence helped solidify cultivation systems that produced reliable results.

The Black Patch Wars represent a case study in inventing tradition. The region emerged from the violence more identifiable as a single unit, producing a unique type of tobacco, than it had entered the twentieth century. There is little evidence that people used the term Black Patch to describe the area, or the dark leaf that some of its tobacco farmers produced, before the Black Patch Wars. A Hopkinsville newspaper from 1896 used the term in a way that made it seem descriptive of part of the cultivation system: "we still love Bettie and the babies, the boys in the furrow and the little black patch."[46] Perhaps locals had used the term, but it appears only rarely in public utterances until about 1904, the year the PPA was formed. The Dark Tobacco District Planters Protection Association likewise founded a regional newspaper called the *Black Patch Journal* in 1907. By 1909, warehouses advertised their facilities to "Black Patch Farmers" in the Hopkinsville newspaper. These were the years in which secret societies proliferated. Night riders, who wore masks, robes, and hoods like the Ku Klux Klan and took people from their homes for whippings, to sabotage their fields and threaten worse, and the activists who terrorized tobacco towns, dynamiting and burning tobacco warehouses, all sent signals to more people than tobacco trust buyers.[47]

Farmers may have thought that they were creating barriers to entry by pooling leaf in cooperative warehouses and allowing only related commodity producers into their activist societies. Their actions did help define their product as unique—limited in its growing region, in its varietal designation, in its market purposes. In fact, however, this resolution of quality uncertainty had results quite the reverse of those intentions: the product became more malleable. Defining the crop according to its characteristics meant that goods possessing those qualities were interchangeable. Tobacco varietals helped commodify the crop: in the Black Patch, the better that growers met the needs of their buyers, the less flexible their production methods and the markets for their goods. This was similar to the result of defining Bright Tobacco as flue-cured leaf. Only the effects of seed, soil, cultivation, and harvesting methods required duplication—not the causes. As curing method solidified into classification, flue-cured tobacco cultivation spread across the Piedmont to the coast, and down into Georgia and Florida. This made it possible for those British settlers in Rhodesia and Nyasaland to begin producing Bright Tobacco—with the aid of empire, of course, and its institutions of preferential pricing and scientific support. Once both buyers and growers agreed on desirable qualities, those characteristics could be reproduced elsewhere. Growers who made crops to satisfy specific purposes found themselves with few buyers, each with precise needs, not competing for the leaf.

FEDERAL ANTITRUST ACTION

As the Black Patch Wars raged, the U.S. government moved forward to break up the ATC trust. Big Tobacco was not the only target. Antitrust action reflected the spirit of the age, and several sectors bore the scrutiny of federal regulators. The suit against tobacco began in 1907, when the government found the licorice paste company that ATC controlled (and with it, the inputs to all sweet navy plug) guilty of violating the 1890 Sherman Anti-Trust Act. The next year, the feds moved against the ATC as a whole. The Department of Commerce and Labor required considerable time and research to trace out the interconnected firms, and eventually the report filled three volumes and twelve hundred pages. In the end, a series of decisions against the combination culminated in May 1911, when the Supreme Court of the United States demanded the dissolution of the firm back into its constituent parts. Buck Duke oversaw the division, as he had overseen the combination. His plan for splitting the enterprise was approved in November 1911, and he retired from the business shortly thereafter. Yet, even as the ATC's structure changed, its strategy continued. In the decade after his retirement, cigarettes finally achieved market dominance—and the ATC had always controlled the production and distribution of that good.[48]

The government-ordered dissolution of the ATC trust did not, however, restore competition. In 1946, the U.S. Supreme Court once again found that cigarette manufacturers "had conspired to restrain trade, had conspired to monopolize, and had monopolized." As in the government's first antitrust actions, the courts of the 1940s showed considerably more interest in the industry's control of distribution and its pricing of finished consumer goods than in its leaf purchasing division. In the first decade of the twentieth century, the government's research showed little interest in raw materials. Of the twelve hundred pages on the structure of the tobacco industry by the commissioner of corporations to the Department of Commerce and Labor, only seventeen scattered pages dealt with leaf at all, and then only as an introduction to the industry. By midcentury, of course, the conditions of leaf cultivation had altered yet again. The crisis of the 1930s had witnessed the federal government's extraordinary intervention in agricultural production. Yet still the courts viewed the actions of the ATC's successor firms in terms of sales, while several independent investigators, economic historians in academic examinations, later found what they considered to be clear collusion among many firms, or what one termed "non-aggressive buying policies."[49]

Buck Duke expressed nothing but regard for farmers' attempts to get better prices for their goods. Nonetheless, his firm's purchasing strategies relied on the segmentation of the market its organization reflected. The ATC bought leaf according to the individual brands made by its constituent firms; the federal government lacked the power or perhaps the will to change that arrangement. In fact, Duke's testimony before the government's antitrust wing assumed that each firm was his competitor, even those within the combination he ran. His grasp of brand-level competition seems considerably firmer than that of his investigators, who thought the price that ATC paid for "goodwill" merely watered the stock.[50] Yet Duke was paying for brand identity, as he described it, "because after one style went out of fashion we would have another style ready for the public to take up." Growers—especially elites—had mimicked this industrial strategy and sought the advantages branding conferred. The organization of farmer resistance according to commodity types only added to the dominance of the manufacturer's segmented market. Ironically, branding agricultural products into varietal types reduced farmers' commercial power. Bright Leaf for cigarettes suited little else. Black Patch tobacco became Italian cigars or snuff, or sold in Africa.[51] Neither type could now find new markets.

ELEMENTS OF TYPES READY FOR REGULATION

At the turn of the century, all the elements that would compose the USDA definitions of tobacco types were settling into place. In the twentieth century, the USDA would classify tobacco types according to their region of origin, their market purposes, and the technologies used to produce them. In this, they borrowed categories established in centuries of trade. The neighborhood in which the tobacco had grown had long served merchants as a descriptor that stood for other characteristics concerning the commodity: Virginia tobacco had recognizable characteristics in world markets, and so did Kentucky leaf and strips. The markets each type served, although negotiable, had well-established relationships and histories. Kentucky tobacco was used by the American Snuff Company, or the Italian *regie*, or specific buyers in Africa, with their established brands and markets. Firms had long worked together to serve both ends of those trade routes. As for technology, methods with extensive histories had begun to creak into new forms, shaped by their changing contexts. These contextual causes included the taxation that segregated agricultural from industrial production and the structures built upon emancipation: credit for financing production on an annual basis and the harvest, curing, and marketing methods that were soon to join environment and demand as the definitions of varietal types.

The rough taxonomy of tobacco types that appeared in the 1880 census proved a useful framework for associating technological choices with market purposes and region of origin. Then, the rise of the tobacco monopoly, with its brand-name consumer products, segmented market, and oligopsonistic purchasing practices, called forth similar actions from farmers, efforts to erect barriers to entry that mirrored the strategies of the manufacturing sector. Farmer protests against their combining buyers took shape in organizations based on the commodities they produced, associating region with technology of production and growing commodities for specific markets. Although these organizations backed political figures and parties that ultimately failed, they succeeded in establishing institutions that provided farmers with market information, advised specific technical approaches to leaf production, and contributed to the farmers' acceptance of types as descriptions of what they grew. Thus, specific types that served well-defined markets and grew in specific locations under established agricultural regimens achieved stability through the changing patterns of market relations and regulations. Helping farmers better meet their markets locked them into distinct technologies of production in specific regions. The institutions did not yet regulate, however. They merely recommended practices.

The protests against taxation with which this chapter began, those objections by the Louisiana tobacco farmers classed as manufacturers based on their methods, demonstrate the difficulty in distinguishing agriculture from manufacturing. They represent one example of the complexity of the process by which farmers accepted type designations as descriptions of what they grew. This process took shape in different ways in each case, for each tobacco type. The case of Bright Tobacco is well known, although (like the Black Patch Wars) the actions around type formation are not recognized as such: the invention of tradition—in cultivation methods, credit and marketing arrangements, and labor organization, which together represent a technological system that defines a tobacco type. These were moments of political organization by farmers in response to changed market structures; they produced leaders and established new institutions that mediated markets on behalf of farmers. Scientific societies and agricultural universities supported the specific tobacco culture that seed sellers, fertilizer companies, and flue vendors developed and spread. As farmers argued that this was the way they had always done it, methods solidified in specific regions and produced tobaccos desired by particular markets. The next step was formal regulation of places and methods of productions into types serving specific markets. Legal designations awaited only crisis to become regulation.

CHAPTER SIX

Stabilization

"Therefore be it resolved by the farmers of Caswell County," declared the Bright Tobacco farmers of the North Carolina Old Belt in August 1933, "that the President of the United States take a hand in the marketing of the tobacco crops." The warehouses had opened on the appointed dates in Georgia and South Carolina, and the Bright Leaf growers found that the prices offered for their crops once again did not pay the cost of producing it. In this environment, with the government at work in so many sectors of the economy, the regions' leaders found new openings to push through their hopes and ideas for government aid and market control. Moreover, they were willing to give up certain freedoms in exchange for help. Viewing overproduction as a problem familiar since colonial times, faced anew with the low prices offered by Big Tobacco (despite the government's antitrust action), "farmers are willing to co-operate with the Administration and reduce the acreage." When the government addressed the problems facing tobacco growers, however, the structures already in place shaped federal efforts. Legislation propped up the existing economic frameworks into obdurate, seemingly natural structures.[1]

As scholars have long been aware, many elements of the New Deal already existed, on the shelf, as it were, proposed and scrutinized during the Progressive Era and the Great War. Shelved during the prosperity bubble of the Roaring Twenties, theories of social betterment and structural reform lay ready to hand when the crisis came. Some of those ideals had already reached fruition, as wartime demographic shifts and antilynching campaigns brought the race relations of the South into a new era. At the same time, urban governments had met the demands of exploding immigrant populations—first, with boss-ism that filled government gaps, and eventually with civil service reforms. Antimonopoly activism, too, had resulted in federal trust-busting and popular resistance to corporate power. For some scholars, the era appeared as a "search for order" in the face of increasing anonymity and the decline of social controls monitored by face-to-face relations. For others, the Progressive Era represents the clearest instance of international cooperation against social decay, a globalized age of ideals and ideas truncated by the First World War. The New Deal's tobacco control programs were among many interwar measures that were adaptations of Progressive Era thinking and planning. These early twentieth-century ideals eventually established an alliance between government and business that persists today.[2]

The Federal Warehouse Act emerged in 1916 from war preparation, for example, and became a predecessor to the New Deal agricultural control programs. Preparing for the war demonstrated "that the farm marketing machinery of this country is seriously weak, insufficient, and inadequate," and Congress believed that a new marketing system would set things right. A "proper relationship between the storage and banking systems of the country" would store crops (not only tobacco, but other produce too) to wait for higher prices. Warehouses would provide receipts for deposited crops that could function as collateral for loans from banks, to be repaid when the stored commodities actually sold. Therefore, farmers could still make money at harvesttime while holding their products from the market until needy buyers had incentives to pay more. Such a system had been a goal of the government institutions that took over the informational work of merchants after the Civil War. Dreams of cooperative marketing persisted, pushed by the land-grant schools and the USDA. The Federal Warehouse Act, however, provided facilities even for individual farmers to receive a harvest payday while holding their crops off the market. The federal warehouses intended to end the volatile price cycles of supply and demand so long ascribed to "overproduction."[3]

The United States became further drawn into the European conflict in the next few years, however. High prices for agricultural goods obtained while the war raged, and tobacco was no exception: dark fired and air-cured leaf rose from six cents a pound in 1915 to twenty cents in 1918 before falling to five in 1920. Bright Tobacco's value traced an even sharper arc, from ten cents to forty-five, dropping in 1920 to the still-high twenty cents a pound. World War I also marked the turning point toward cigarettes as the primary tobacco ingestion method. From U.S. soldiers to respectable Canadian ladies, smok-

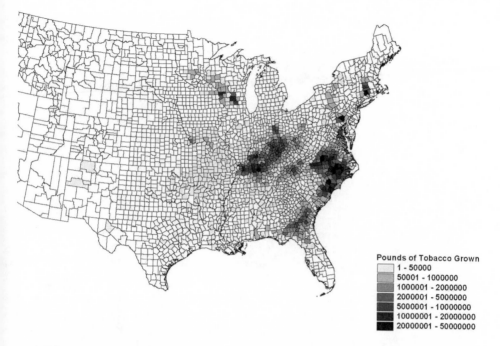

Pounds of Tobacco Grown
1 - 50000
50001 - 1000000
1000001 - 2000000
2000001 - 5000000
5000001 - 10000000
10000001 - 20000000
20000001 - 50000000

Tobacco production intensified in the twentieth century. While tobacco agriculture moved into new locations, established regions of origin also grew more leaf, compared with the census of 1900 (see p. 116).
Source: U.S. Census for 1920.

ing took hold during and after the war. Agricultural prices therefore stayed at wartime averages throughout the 1920s, despite fickle years. When the drop came, in 1930, it was steep. Although flue-cured consumption increased rapidly during the 1920s, still the quantity produced each year outstripped demand by about 50 million pounds. Crop spread doubled the acreage under cultivation, and adopting the type meant capital investments in new technology—curing barns especially. In 1930, "equipment was available to produce approximately seventy-five percent more flue-cured tobacco than in 1920." Demand and supply cycles familiar from colonial settlement still bedeviled twentieth-century tobacco farmers.[4]

Farmers seeking better prices and more market control first tried to help themselves—albeit aided by the models and government support of scientific agriculture. In 1921 and 1922, with the encouragement of the state schools and the various agencies of the USDA, Bright Tobacco farmers of Southside Vir-

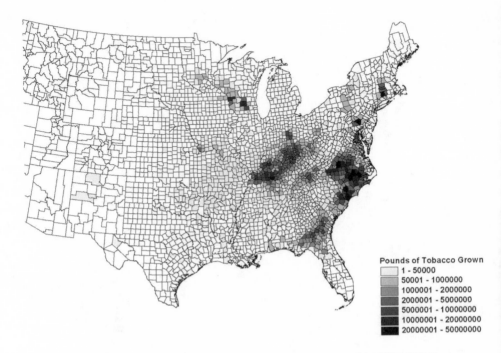

Pounds of Tobacco Grown
- 1 - 50000
- 50001 - 1000000
- 1000001 - 2000000
- 2000001 - 5000000
- 5000001 - 10000000
- 10000001 - 20000000
- 20000001 - 50000000

When agricultural prices dropped during the 1920s, farmers pushed the federal government to establish production controls and price supports. These allotments limited tobacco to the quantities grown in 1930. Thus, the federal programs locked tobacco agriculture, the types produced and where they grew, into this pattern. While many scholars think that particular tobacco types grow only in specific regions, Bright Tobacco agriculture was still spreading when the New Deal programs shut down the expansion of the culture.
Source: U.S. Census for 1930.

———————

ginia and the North Carolina Piedmont attempted once again to meet their buyers with the old weapon of cooperative marketing. This time, however, these farmers had powerful institutions on their side. They tried to reorganize their crop's marketing structures with the Tri-State Growers' Cooperative but failed; the Cooperative entered receivership by 1926. However, both state and federal governments had begun encouraging such efforts, as in the Federal Warehouse Act that had been established in 1916. Extension agents and faculties from the agricultural colleges facilitated cooperative ventures among farmers. They wanted to help farmers organize, to force the hands of the buyers who needed their products. For tobacco, the effort failed: the long delay between purchase and use, the years in which the leaf sat in giant brick warehouses, prized into hogsheads, sweating with the seasons, meant that dealers and manufacturers often had large stocks on hand. Farmers holding tobacco crops off the market faced the considerable ability of their buyers to wait them out.[5]

The 1916 Federal Warehouse Act had provided what farmers had long sought: places to store their crops and wait for prices to rise. These public warehouses had familiar problems, however. The years of leaf storage benefitted dealers and manufacturers. They kept large stocks on hand and could weather a storm of high prices for farm goods better than farmers could bear the troughs, when prices dipped near or below the costs of production. Manufacturers' grades, too, allowed them to use old and new crops for one purpose; they stocked up in the flush years and waited out any high prices that scarcity wrought.[6] Meanwhile, as farmers had learned in the Black Patch, keeping leaf from its buyers only worked if most of a particular crop of tobacco, leaf with characteristics desired for a specific purpose, stayed off the market all together. This required identification of the crops. So did using receipts as collateral for the leaf stored in federal warehouses. At first the Act intended to "keep separate" the crops of each grower "so as to permit at all times the identification and redelivery of the agricultural products deposited," but insuring stored crops against fire meant establishing their value. Crops had to be interchangeable, which made standard type and grade designations more necessary. Revenuers, too, wanted to use such information to generate statistics.[7]

In Elias Carr's neighborhood, for example, in East Carolina, where Bright Tobacco culture had spread toward the coast around 1890, the regularity of technical methods facilitated the production of Bright Leaf tobacco, and the use of government graders contributed to the continuing viability of the region's leaf. Tentative USDA grades appeared in 1925, with specifications concerning size, color, body, and weight, "elements of quality that frequently determine the use to which it is best suited." Moreover, as always, the regularization of production affected the technological system in ways that reflected the organization of agricultural labor, and government grading could provide means of judging work:

It is easy for me to tell what tenant makes good tobacco year in and year out, no matter what price he receives. The price does not show that he is a real tobacco farmer, but the Government's grades of each crop that he grows does. . . . The fact that each tenant knows that his crop will be Government graded makes him more careful in the curing and sorting of his crop. The fact that I have the same tenants all the time makes the above information very valuable. When I get a good tenant he doesn't move each year.[8]

The receipts supplied by the Federal Warehouse Act finally necessitated the designation of tobacco by type, and the USDA stepped in to classify the leaf. Informational work once performed by merchants had become the preserve of government science, which drew upon existing categories and elements of descriptions of the leaf for sale—categories developed over centuries of trade—to develop its legal names. In order to state "the types of tobacco" on warehouse receipts (and to insure the warehouse contents), the government needed to develop a formal classification system. It found, however, that "there was no classification of all types of tobacco which was commonly understood." So investigators researched the names that farmers used to describe their product in different regions, the seeds they employed and the techniques they applied to produce it, and the world and domestic markets that demanded their crop's preparation in particular ways. In 1925, the USDA reduced 315 tobacco types into the six principal classes known today, three curing methods and three cigar parts. The twenty-nine total types within these classes received numbers, so that U.S. tobacco type 11 was Old-Belt Flue-Cured, along with a whole paragraph of names it had once been called. New-Belt Flue-Cured was type 12 or 13, depending on its coastal growing location. Region of origin and market purposes, borrowed from merchant descriptions, combined with technology of production to define the article.[9]

For this was the very definition of a commodity: a "standardized good, which is traded in bulk, and whose units are interchangeable." This definition was the point to which all this history had led. In each period of American history, the technological system of agricultural production had driven tobacco toward further specificity of crop characteristics. The technologies of producing those characteristics, however, took different forms in specific times and places. From the inspection laws of colonial Virginia's lawmakers and their efforts to raise the value of their staple sprang a production system that limited the production of second growths, ratoons, and suckers. This maximized the cultivation of first-growth leaves that would be harvested all at once, the whole stalk, as frost threatened or other crops ripened, or when the leaves looked perfect for the next intended purposes. The peak labor demand at harvesttime in the Chesapeake system of tobacco cultivation escalated a colonial system of slave labor later increased by cotton, the tobacco

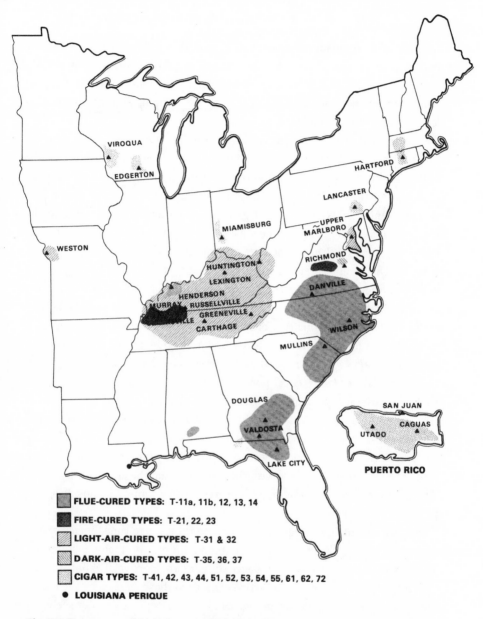

FLUE-CURED TYPES: T-11a, 11b, 12, 13, 14

FIRE-CURED TYPES: T-21, 22, 23

LIGHT-AIR-CURED TYPES: T-31 & 32

DARK-AIR-CURED TYPES: T-35, 36, 37

CIGAR TYPES: T-41, 42, 43, 44, 51, 52, 53, 54, 55, 61, 62, 72

● LOUISIANA PERIQUE

The U.S. Department of Agriculture established the tobacco types in the 1920s, in order to ensure the crops held according to the Federal Warehouses Act of 1916. The types were defined along three axes: (1) where they were from, (2) what they were for, and (3) the technology used to produce them. Curing methods distinguished one tobacco variety from another, rather than seeds, genetics, and nature.

Source: "Tobacco in the United States," USDA Miscellaneous Publication No. 867 (GPO, 1979), 5.

manufacturers' demand, and westward expansion's supply. Then, in the ante-bellum years, tobacco types proliferated as the market (both supply and de-mand) expanded. That some buyers intended export while others worked at home only further solidified divisions between shipping and manufacturing leaf first initiated by the inspection laws.[10]

The dramatic and wrenching changes of the Civil War brought the tobacco types into being. Taxation segregated agricultural from industrial processes by levying taxes on manufactured goods only, defining these first by pro-cesses and later by limiting sales, licensing each node in the commodity's web and limiting flexibility all along the chain. Agriculture and manufactur-ing became distinct sectors: one was taxed, the other was not. Emancipa-tion changed agricultural practices, the harvest and curing and marketing methods that would later define most tobacco types. Credit arrangements enabled agricultural production under an emerging system of paid labor and contributed to the tuning of the technological system, establishing new methods at the same time that crop spread made regular techniques matter more. Various forms of sorting, curing, and packaging for sale became locked into place as agricultural, not industrial, processes, and these combined with region of origin and market purposes—the old-time merchant information—to define varietal types. Agricultural science, too, as a branch of the federal government newly empowered by the Civil War, became institutionalized in state schools, agricultural societies, and of course the USDA and its extension agents. This federal support of tobacco (and other) farmers reached its ze-nith in the twentieth century, with the New Deal and the entire bureaucracy devoted to the preservation of agriculture by price supports, subsidies, and production allotments.[11]

Just to cement the recognition that these things take time, however, it is important to note that the process of commodification remains incomplete. There is no futures market in tobacco, despite an abortive attempt in 1934. Despite careful grade and type distinctions, even those devised by the indus-try to describe its own needs, no one can bid a leaf's price without seeing it, touching it, examining it. As centuries of tobacco marketing demonstrated, the more specific the needs of the buyer, the less interchangeable the articles he purchased. Since the manufacturers had no obligations to abide by the grades of the government, the use of disinterested graders fell by the way-side. Utilizing receipts as negotiable instruments never quite took hold after the Great War, unless the receipt corresponded exactly with the tobacco to which it referred. Tobacco was not yet, even after centuries of production, a commodity in the strict sense of formal economics: a good interchangeable within its classification system and quality designations. Sweet crude oil priced by the barrel is assumed the same across the grade. As William Cronon saw, the grain elevators of Chicago implied a similarity of wheat within types and grades that farmers produced and buyers accepted. No such institution yet classifies tobacco in ways its buyers accept.[12]

HOW CRISIS HELPED

The economic crisis had afflicted farmers longer than the rest of the nation, like canaries in the coal mine for the structural problems unseating the basis of American economic growth. When the crisis reached other sectors, it became possible to act. The election of Franklin Delano Roosevelt may have come about simply as a revulsion against what had gone before, the monetary contractions of the Herbert Hoover years and a sense that the president viewed struggling Americans as whiners who should bear the brunt of the market's correction. The election provided FDR with significant popular support, however, and from his first hundred days through the constant tweaking of policies and agencies in the decade that followed, the New Deal did in fact transform the domestic economy in ways far-reaching and still felt today. It provided the social safety net on which the U.S. worker relies, as do employers who benefit from government enforcement of paycheck contributions for retirement. The institutions that crisis enacted became themselves shaping mechanisms for economic behavior. Yet however much the New Deal transformed the American economy, it also drew upon existing industrial forms, solidifying them into structural forces.[13]

The relationship between the tobacco industry and tobacco agriculture provides a vivid example of the New Deal's conservative effects. Suggestions for ameliorating the problems of tobacco farmers took familiar forms. The Chamber of Commerce of Kinston, North Carolina, for example, suggested that the "Government" should plan on "preventing the sale of scrap and very low grades of tobacco and relieve the pressure on the market by taking off the market those types and grades of tobacco of which there is a surplus." Since the 1600s, of course, this had been one strategy Virginia's Burgesses had tried to boost the price of the commodity—preventing the sale of poor-quality parts of the plant. Movement in that direction had been accomplished in the 1730s, with the inspection laws that worked; now, surplus types and grades were as bad as scrap and seconds. Distinguishing good from bad tobacco, however one might draw the line between them, still required government inspectors and the agreement to such categories by the entire commodity chain. Unsurprisingly, in the twentieth century as in the seventeenth, neither could entirely encompass the market. By the 1930s, however, a long history guided market regulation. It thus looked more natural and therefore stuck better to the tobacco trade.[14]

MARKET REGULATION

The New Dealers sat down with every branch of the industry. Tobacco manufacturers, boards of trade and dealers' associations, warehousemen, landlords, tenants, and sharecroppers all provided input into the programs the government eventually adopted. However, each tobacco type had its own industrial structure and therefore its own sets of representatives. When investigators classified the twenty-five-odd types into "four or five large groups on the basis of competition and uses," they learned that the "price situation of each group is different from that of any other group." From Connecticut, Florida, and Kentucky to Virginia and North Carolina, every region had grown its own business, its own version of the weed. Each produced a unique commodity serving particular purposes, and each tobacco type made its way to market through established channels that the government had no intent to disturb. Each individual tobacco type received separate attention from the Agricultural Adjustment Administration (AAA). Established in April 1933, the AAA began considering plans for tobacco production controls based on existing businesses. Tobacco specialists relied on established systems of leaf cultivation, agricultural financing, and marketing—even for those arrangements that had solidified only after the Civil War. The production controls for Bright Tobacco also depended on the quantities that individual farmers were already producing.[15]

Each industrial structure, each commodity chain, sent its own representatives to the government regulators. Because they first sat down to address the commodity producers' problems during the midsummer cultivation season, and because the government intended to use the harvest as the basis for planning the next year's production quotas, the administration had some time to devise methods. The agricultural planners therefore dealt with the different types of tobacco in an order determined by the price situation for each type. For example, since the price of cigar tobacco was half "the fair exchange value," while Bright Tobacco's price rested nearer 80 percent of its value, the AAA's production division dealt with the cigar tobacco types first.

When the AAA turned its attention to flue-cured tobacco, it reified a structure that had long been building. Small changes were necessary, as we shall see, to tune production to the emerging twentieth-century political economy. These wrenched less, however, than the shifts forced by emancipation half a century earlier. Still, the methods of tobacco harvesting, curing, and marketing and the credit system that underlay landlord-tenant farm output nonetheless moved a little, solidifying into regular, accepted technological systems of production, to meet the changed conditions that economic crisis and government response had initiated.[16]

Marketing methods for Bright Tobacco, as for the other types, had settled into a routine by the 1930s. Marketing boards and warehousemen therefore advised the government in developing its plans. In July 1933, the local boards of twenty-nine tobacco towns received instructions to send to the AAA "several copies of the rules and by-laws of your organization, including the scale of warehouse charges in effect during the last season." From Danville, Virginia, to the Black Patch towns of Hopkinsville, Kentucky, and Clarksville, Tennessee, local boards of trade submitted their standards. Sixty percent of the flue-cured tobacco stayed home, an input for domestic manufacturing. For that reason, government could take its cues from established market structures. The New Dealers also worked hard in the tobacco system to maintain the sharecropping and tenancy arrangements that had shaped each agricultural calendar and dominated production methods after the Civil War. They wanted to ensure that no one was thrown out of work as a result of their programs. As a result, the allotments that government agencies granted and the checks they wrote paid tenants as well as landowners. They preserved, as if natural, the various tobacco cultures that had emerged after the Civil War.[17]

For Bright Tobacco, the calendar of production received official attention. For example, "the opening date for the South Carolina markets" became the subject of intense lobbying by "several interested parties," as the AAA considered opening the market a week earlier than scheduled (on August 8, 1933, rather than August 15). The 1933 opening dates of all the Bright Tobacco markets had been set by the Tobacco Association of the United States during midsummer deliberations. This association had representatives from all the principal markets of Virginia and the Carolinas, as well as a few from Florida, Georgia, and Kentucky. The Association set opening dates for sales from Georgia through the Old Belt of Carolina-Virginia and met again in July to reconsider the schedule, but it decided to stick with its first decisions. Some people complained that fixing the opening date of various markets was un-fair: "it is done at a meeting of the warehousemen and manufacturers without any representation of the farmers." The Tobacco Association, however, urged the government to stick with the Association's schedule. Opening day helped buyers schedule and staff, of course, and "any change now would upset the whole schedule, and I hope this matter has been adjusted in such a way that there will be no further discussion about it."[18]

Farmers wanted more, maybe even an immediate subsidy to raise the price of leaf as it came to market that very fall, funded perhaps by a processing tax. As opening day approached the markets of the Bright Tobacco growing re-gions, from Georgia and South Carolina to the border and into the New Belt of North Carolina, farmers in Elias Carr's neighborhood kept their eyes on the price. Disappointed in the ten to twelve cents a pound Bright Leaf brought to coastal growers, the North Carolina farmers demanded of (and received from) their governor a "voluntary marketing holiday" that, with the cooperation of the warehousemen, spread to South Carolina and awaited the action of the federal government. The plan was not long in coming: guaranteed leaf prices at parity with those received in 1914! If buyers offered a lower price than that tendered in 1914, the government would make up the difference. In exchange for this price guarantee for their 1933 crops, farmers had to sign agreements limiting their production the following year. The farmers embraced govern-ment aid and government control so wholeheartedly that by the time "the sign-up" closed, 95 percent of the North Carolina tobacco growers had signed contracts to reduce their crops the following year.[19]

In 1934, farmers limited their tobacco production based on quantities grown in the three previous years. By paying both landlords and tenants for their crops and insisting that tenants displaced by allotments be replaced with new workers, the government preserved the sharecropping system. The New Deal cotton program did not work the same way; according to Pete Dan-iel, "federal agricultural programs drove workers from the land" in a process he termed "the southern enclosure." Cotton farms grew in size, tractor use increased, and "wage laborers supplanted other tenure classes." The federal government caused significant shifts in the cotton culture and in Southern society more generally. In contrast, however, equivalent federal programs in tobacco production reflected and solidified existing agricultural structures, from farm size and labor arrangements to cultivation methods and market-ing traditions. By establishing allotments based on quantities grown between 1931 and 1933, according to Anthony Badger, the government "tended to freeze agriculture into the pattern of those years." Indeed, the agricultural pattern, the technological system that made tobacco bright, reached closure under the legislative enactments that told farmers not only how much they could grow but of what kind of tobacco.[20]

CLOSURE

The New Deal brought into closure the technologies that produced each tobacco type. Many elements of Bright Tobacco's cultivation system were already solid, of course—topping and suckering, for example, had developed as part of the Chesapeake system of tobacco production during the colonial period. On the other hand, hilling had begun to seem unnecessary in the twentieth century. Other parts of the technological system were also still developing, especially those tasks related to harvest, curing, and marketing. Priming was more common a harvesting method than it had been a generation before but was not yet universal. New Deal oral historians recorded the farmer who had not started priming "until about 1919," and some took up the practice even later. "During the first quarter of this century, tobacco was harvested by cutting the whole stalk . . . successive strippings of leaves did not become prevalent until about 1930," some historians later discovered.[21] In fact, the New Dealers encountered obstacles whenever they tried to use priming to limit the quantities produced. They found that "only a part of the flue-cured crop is harvested by priming," so "to leave a portion of the plant unharvested" was an impractical method of controlling production. As priming took hold, New Deal policies both shifted technologies and crystallized them into standard methods.[22]

Yet, by engaging every sector interested in the production and use of each tobacco variety, the federal laws locked into place the industrial structure that spread out from and supported each one's production. Old blurry boundaries among the sectors had meant that antebellum farmers and dealers sometimes became manufacturers, but taxation had limited who could sell to whom, establishing boundaries between agricultural and industrial production. Agriculture had also changed after the war: emancipation had transformed so thoroughly the value of labor and, with it, the cost calculations of producing the crop that an entirely new technological system of harvest, curing, and marketing eventually emerged to structure the growth and sale of the antique staple. Human choices made this history happen, however. Structures do not change on their own. Elias Carr turned merchant for his tenants, experimented with flue-curing, led the Farmers' Alliance in limiting membership to farmers, and became the Democratic governor of North Carolina. The Gravely family sued for the right to name their products after Henry County. Flappers smoked cigarettes. Sprigg Russell signed a contract in 1865 that ended his obligations to his former owner by the New Year. Although New Deal legislation solidified these relationships, they had first been constructed by individuals.

Thus, the production controls and price designations drew on market relations and the categories such negotiations and transactions had established, making these categories the subject of law. Any flexibility between buyers and sellers in the marketplace was now something forbidden, gone and forgotten. A farmer who grew flue-cured tobacco in the quantities specified by his federally granted allotment could no longer decide at the last minute that air-curing might best serve the needs of the market, or the characteristics of that year's crop. Moreover, flue-curing his tobacco categorized it as belonging to a particular USDA class of produce. If he happened to open his barn doors to find his tobacco brown, or reddish, or even "blue as indigo," it was still Bright Tobacco—just of poor quality, as evinced by its color. No merchant would say it was "good dry Shipping Tobacco's & very sweet" or "of a Manufacturing quality" or classify it as "French Tobacco," even if these descriptions were accurate. The categories established by the government dictated that tobacco from a particular region, produced by the flue-curing technology, served particular markets and none other. The system was closed. Variation had become impossible.[23]

THE TWIST

None of it was true, however. The government was just about to learn that its categories were fictional, although long established in trade and now limiting in the relationship between buyers and sellers. As the earnest agricultural experts and federal legislators and advisors examined the tobacco trade type by type and established production controls on farmers that dictated not only how much but what kind of leaf they could grow, other government scientists dropped a bomb on the whole enterprise. Genetics had finally caught up with the industry, and the first tests of tobacco shocked the investigators. Tobacco types of such different colors and characteristics as to fall into entirely different USDA classes were "so similar as to be almost indistinguishable," both from one another and from the Orinoco first exported from Jamestown in 1617. As the investigators established, when the cultivation of dark tobacco had begun to give way to bright cigarette tobacco, for example, "[t]here was no significant change in the varieties of seed used." Moreover, "the present strains of the Orinoco as a whole do not differ greatly from those employed originally." The geneticists sounded as if they had been fooled by the seed sellers into thinking that more varieties existed than their biochemical analysis could support.[24] They realized that technology, not genetics, defines the commodity.

Twenty-first-century genetics has only added further emphasis to the point. Even burley and bright, which the 1937 geneticists saw as inheriting two separate lines, are only a little different from one another. The twenty-first century has witnessed an explosion of genetic knowledge, from the Human Genome Project to the reinvigoration of ideas about race for medical purposes, as certain medications appear to be more effective on certain related populations, classified according to their racial makeup. Even in these explorations, however, as the most modern forms of gene sequencing acquire deeper and more detailed information concerning tobacco types and varieties, the similarities among them appear more obvious. In 2007, a group of geneticists writing in the academic journal *Annals of Applied Biology* attempted further assessments of the genetic diversity within and among tobacco types. Their investigation was principally devoted to devising new experiments and methods of analysis. To the extent that there are differences among the cultivated tobaccos, the geneticists discovered, the distances between genetic markers within the genome do fall into categories based on "manufacturing quality traits." The abstract of their conclusions, however, maintains and extends the surprise of the 1937 findings.[25]

Genetic differences do mark some distinctions among the seeds, but, even in 2007, scientists seemed surprised by how little of the difference the seeds could explain. Twentieth-century breeders served commodity production by importing distant strains and creating cultivars for very specific soils and disease environments. This specificity is possible as the crop has stopped spreading to new locations, but it is partly necessitated by the disease environments that develop from growing the plant in the same fields over long periods of time. Genetic differences are not nearly powerful enough, however, to explain the morphological, appearance-based distinctions among tobacco types. The types of tobacco that sell in the marketplace for different purposes have been created specifically for those purposes, and the technologies employed in those creations developed within unique, specific, historical frameworks. These frameworks, as well as the different technological systems for producing tobacco leaf that each contains, are the products of human choices. Technology crystallizes its context: economic relations and market regulation, and the entire structure of incentives for particular agricultural production techniques and industrial relations. Methods and machines embody social standards and cultural assumptions. Cemented into place by the New Deal regulatory structure, constructed within the context of changing economic relationships and institutional frameworks, with aid from government regulation and scientific investigation, the techniques of harvesting tobacco by priming, curing in flue barns, and selling the crop by auction in public warehouses can be counted on to make tobacco bright.

At last, with the stabilization of regulation, the twentieth century had come into its own. The post–New Deal world was indeed a consumers' republic. The power of branding had never been stronger, which is why so many scholars view it as new in the new century. From the Camel cigarette, introduced around World War I, to the Marlboro Man, who first appeared after World War II, the cigarette method of ingestion brought tobacco manufacturing and, with it, tobacco agriculture into a new phase. The cigarette century relied on Bright Tobacco. When the industry consolidated in the late nineteenth century, organized into bigger units, the agricultural units became smaller. The New Deal allotments, production controls, and price subsidies kept this scale in place. For several decades, farming Bright Tobacco remained a family affair, with small allotments assigned to farmers who worked the land in ways solidified after the Civil War. Big Tobacco, the global behemoth, likewise emerged in the postbellum era. Neither the agricultural nor the industrial sector exists outside of human control. Each developed in a specific time and place, as a result of institutional frameworks into which it fit as if it had emerged without human effort, a product of the plant, as natural as rain. So it goes. When humans build something that works, it looks natural. It becomes resistant to change.

MECHANIZING THE HARVEST

It seemed as if it had always been this way: small family farms provided the raw materials for a big mechanized multinational industry. Because the cultivation system looked so natural, it was hard to change. It proved nearly impossible to mechanize the harvest—to replace by machines the most labor-intensive moment of agricultural production. Although priming had solidified as a harvest method only in the twentieth century, it became after the New Deal simply the way Bright Tobacco was harvested, as if the technique lay outside human control. Priming fit well with the postbellum calendar of tobacco production, so abbreviated from colonial days and antebellum marketing structures, because it made some of the work of grading the leaves (according to size, based on their position on the stalk) take place during the harvest. Then, because each barn full of leaves underwent the same temperature variations and produced leaves of about the same color, all the sensible indications of leaf quality seemed the same. This got the crop more quickly to market. Stripping no longer dragged through the spring, while the tasks of marketing facilitated rapid sales. Priming worked. It fit into the technological system of producing Bright Leaf for cigarette manufacturers, a system of agricultural production initiated by emancipation and stabilized in the decades after the war.

Technology seems the natural response to material demands, the one best way to get the job done. Yet it is instead the result of human choices made within economic and institutional frameworks. The methods of producing tobacco changed in different times and places, in different political economies. Once the technology reached closure, however, the methods defined the types. They could not be changed—not without transformations to the entire world within which they had taken shape. For this reason, mechanization of the Bright Tobacco harvest did not occur until the 1970s. This slow adoption of harvesting machines provides the answers to the questions this book initially posed. Why did the tobacco industry mechanize while agriculture became smaller in scale, more artisanal in its skills and regional in its production? Why did agriculture resist for so long the scale-up in its production? The reason was the technology: the harvest, curing, and marketing methods that were themselves created in the postbellum context and then solidified by New Deal laws. The whole system had to change for the machine to work. If Bright Tobacco was natural, from the seed, then harvest and curing techniques were only revealing the plant's essential characteristics, and machines had to reproduce their effects. However, the structure of cultivation systems limited production in ways that prevented profitable adoption of machines.

Answering the questions posed at the start of research requires understanding that producing Bright Tobacco included not just cultivation, harvest, curing, and marketing methods. The technological system encompassed all the social structures and cultural practices of the people who grew the plant, as well as the regulated relationship between supply and demand. Picking individual leaves as they ripened, in barn-filling quantities, spread the harvest over several weeks and made good use of extended families—as in the farmers who remembered that "we barned Uncle Montgomery on Monday. Saturday was Uncle Dewey's day. All of us worked together." Families lived near one another and took turns in one another's fields, drawing on more labor than the immediate workforce, rotating workers through the fields, getting every crop on the place harvested, barned, and cured, in its turn. Just as the harvest method of the seventeenth century had contributed to the stability of a system that employed slave labor, the harvest method of the twentieth century drew upon and strengthened the stability of farm families over generations. Gender and age divisions of labor bestowed sense on the system, too: women tied leaf onto sticks while children helped their parents, and men both plucked the leaves and lifted laden sticks to fill the flue-equipped barns.[26]

The task complex of harvest, curing, and marketing had become the definition of the type. The technological system was so skilled and small-scale and regionally specific, so deeply tied into its context, its regulatory and customary environment, that it could not change without a corresponding shift in the social and legal structures of tobacco cultivation. Few efforts were made. All the way up until 1950, only two attempts to mechanize any part of tobacco agriculture received patents. Both of these machines tried to move the work of stringing leaves onto sticks out into the fields. Either could have revolutionized the harvesting process had it been widely accepted, but both relocated skilled women's work out into the fields and required that looping be done at the same time as harvesting. Increasing the work of men and making less use of available women did not fit farm size or twentieth-century labor arrangements. Poorly strung tobacco also posed dangers at curing time: one leaf inadequately strung might drop from the stick onto a hot flue below, sending the barn filled with perhaps a quarter or a third of the year's harvest up in flames. Unsurprisingly, neither of these machines took hold, because the labor organization of the harvesting methods did not fit their vision.[27]

For the postwar generation, however, the technological system of tobacco cultivation changed quickly. In the two decades after 1950, forty-four devices to speed up bits of the Bright Tobacco culture received patents. Some made an effort to reorganize old forms of labor into new locations.[28] Other machines tried to mechanize priming itself. Their inventors assumed regularity in ripening and thereby mechanized the judgment of pickers.[29] Most bruised or tore the leaves, which damaged their market value. Some resulted in lower yields because they failed to prime every leaf, while farmers working under a system of acreage allotment likely wrung every leaf from their acres. Machines cost money, too. Cost calculations took into account the available

The twentieth-century system of producing Bright Tobacco made good use of family labor, with different tasks assigned to different family members. While men harvested individual leaves, women looped the leaf onto sticks for placement in the curing barn, where wood fed the fires that cured the leaves to their bright color. The priming technique of harvesting employed extended families like these, who rotated their labor through the fields, taking individual leaves as they ripened.
Source: "Looping tobacco near Wilson, NC ca. 1926," North Carolina Collection, University of North Carolina Library at Chapel Hill.

labor and also had to recognize that profit margins were slim, even with government support. Tenants had little incentive to invest capital in bulky expensive machines. Landowners who operated under allotments limited by the government may have feared that debt would jeopardize their property. Such dangers lurked in every improvement, even in those less technologically forward: one corn farmer remembered that large landowners had sold smaller proprietors mules so poor "the wind could blow them down" and thereby acquired both their land and their labor.[30]

In the end, mechanizing the Bright Tobacco harvest required changes to the whole political economy of its culture. The generation of farm-boy veterans who went to college on the G.I. Bill learned modern ideas of efficiency from the agricultural colleges. Extension agents and agricultural schools advocated new techniques to replace cultivation tasks with machines and chemicals. Chemicals sped the preparation of seedbeds, while machines pulled workers four abreast through the fields, seated low to the ground, transplanting seedlings with cones to poke holes and through which to drop the plant into place. A topping machine, patented in 1954 and improved in 1969, took off the flowering tops of the tobacco plants and sprayed the stump with a formula that prevented the growth of suckers, those second growths that took energy from the commodity leaves.[31] Farmers who had once contracted workers by the year, to top and sucker through the summer, could now need workers only for the harvest. In turn, laborers hired only at harvesttime found other work. Some left the land, and some were driven from it. Landlords could struggle to find tenants, many of whom fled the land for new opportunities during both world wars and again in the late 1950s. With technological momentum developing across the complex of cultivation tasks, harvesting looked like a bottleneck in the production process rather than its culmination.[32]

Inventions that changed the tasks associated with curing eventually spurred the mechanization of the harvest. Some machines still attempted to replace the work of women, the looping (or tying) of leaves onto sticks before curing. One trailer instead pressed leaves between two sticks. Another had a rack on which to simply spike the leaves rather than sending them to the barns for stringing. A trailer designed to hold sticks loaded with tobacco in some unspecified fashion also prevented "rehandling of the tobacco prior to . . . curing." In fact, a cover or "envelope" (patented separately) meant that tobacco could be cured without removal from that trailer. Curing had itself begun to change, but several steps marked the path. First, barns equipped with oil-burning furnaces began to appear. Landlords provided these, as they had the traditional flue-curing barns on farms they rented. Unlike the traditional curing process, which used wood collected by tenants in slack moments, year-round, these new furnaces required store-bought fuel. Even flues became unnecessary as barns came equipped with several furnaces, which kept the heat spread evenly through the leaves. Dials and gauges on the furnaces controlled the heat over the entire curing process and let the farmer sleep through the night rather than dashing up to read temperatures and stoke fires.[33]

Legislative changes to marketing practices helped ease in the practice of bulk curing. Bulk curing meant that the leaves, once primed from the plant, would go into mesh trays for loading directly into the barn. Bulk barns proved popular among farmers. The mesh boxes would release no leaf to drop onto a flue and start a fire. The engineers from North Carolina State also liked bulk barns for their "greater control of moisture and drying conditions" and believed that the system would "accommodate wider ranges in leaf maturity and harvesting conditions (turgid vs. wilted leaves)" than conventional flue barns. The faculty saw the connections between changes in curing and changes in harvesting a decade before a successful harvester appeared, seeing "a breakthrough in simplifying the task of integrating mechanical harvesting with the curing operation. Thus not only is bulk curing compatible with present methods of harvesting, but it ties in perfectly with improved harvesting methods for tomorrow." Economic studies supported these claims, comparing costs and labor across several production systems and making use of various configurations of new and old technology. Finally, federal laws facilitated technological change when, in 1968, they allowed the sale of loose tobacco, changing a practice that had survived since colonial times. Doing away with old methods of grading and tying into hands made bulk curing practical.[34]

Yet the volume of output shaped the ability of farmers to afford any large machine, and crop sizes had been limited since farmers first signed up with the government in 1933. In 1961 and 1962, the government began to permit the lease or transfer of allotments from one farm to another as long as both farms stood within the same county.[35] This allowed farmers to acquire larger parcels and change their cultivation systems. A new generation of boosters also promoted new processes. Chemical companies that manufactured fertilizers, pest eliminators, and sucker inhibitors filled the advertisement pages and features of *The Flue Cured Tobacco Farmer*, a magazine founded in 1964. The magazine allowed subscriptions only to farmers with government allotments of five acres or larger and required copies of allotment certificates along with the three-year, $2.50 subscription payment. Smaller farmers were welcome to subscribe if they paid twice the price. This requirement meant that the magazine reached a very specific audience. Once paid, the subscription kept coming for three years, proclaiming labor shortages and trumpeting the benefits of expensive chemicals, machinery, and gauges to control curing. Machines also freed the farmer from reliance on the dwindling rural workforce of the South. The advertisements for new technology promised a family agriculture without messy reliance on tenants, hired help, or extended kinship networks.[36]

Thus, the legal and cultural stage, the institutional framework, was set for successful mechanization of priming. In 1961, Robert Wilson (a different Robert Wilson, obviously, than the one who packed together the crops of his neighbors in 1785) patented a machine that set rubber defoliators at an angle, and the height of those defoliators could be adjusted for successive rounds of priming as ripeness crept up the plant. He served on the agricultural engineering faculty of North Carolina State University and took the job in order to help people—or so he told his mother, who objected that tobacco was a nasty crop. Throughout the 1960s, with the new farm sizes and bulk curing beginning to shape the possibility for a new technological system of Bright Tobacco production, the design of harvesting machines and elements of harvest-and-curing systems proceeded at North Carolina State. The machine that really worked, however, was memorialized in two patents, granted in 1971 and 1974 (the second version adopted the rubber defoliators from Wilson's 1961 machine). The harvesting machine had been developed under a grant from the R. J. Reynolds tobacco manufacturing company, and it went immediately into large-scale production.[37]

Economic structures—farm size, labor costs and availability, market demand—all changed around the adoption of the new machine. Some of these caused technological change; others were its effects. Government institutions played a role, as laws had limited farm size and regulated who could sell to whom. The harvesting machine designed for R. J. Reynolds broke even at forty to fifty acres. Some farmers who leased their neighbors' allotments could acquire this much land, mechanize not only cultivation tasks but also harvesting and curing, and maybe even free their operations from reliance on paid labor.[38] Even the first attempt to mechanize the harvest had claimed that it saved labor and made it "possible to harvest the tobacco with the same amount of labor as is required to plant and produce it." In 1972, less than 2 percent of the Bright Tobacco crop was harvested mechanically. Within a decade, that figure rose to almost 50 percent.[39] Mechanization accelerated the process of farm enlargement, and the average acreage per farm of Bright Tobacco jumped from 5.2 acres in 1964 to 12.2 in 1978. Increasing farm size helped farmers adopt the machine, and adopting the machine then enlarged farms. As is typical in the history of technology, the structural changes that made mechanization possible acquired yet more momentum from the transformations they had caused.

In this way, the machine designed to harvest Bright Tobacco stood between cause and effect. Technologies accelerate the forces that bring them into being. This middle position helps explain why historians struggle to incorporate

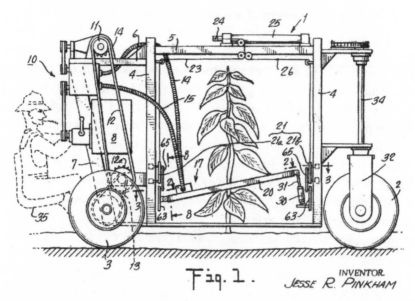

Fig. 1.   JESSE R. PINKHAM
INVENTOR

The priming method of harvesting one leaf at a time became so characteristic of Bright Tobacco production in the twentieth century that it prevented mechanization of the harvest until the 1970s. The harvesting machine that worked picked all the leaves below a certain, adjustable point on the plant. It cost so much money, however, that the whole structure of farm size and labor forces had to change once again, to make mechanization profitable. *Source*: U.S. Patent 3,693,064, granted Sept. 7, 1971.

the history of technology into their standard stories. It is relatively easy to recognize the effects of technological change; its causes are more various and difficult to identify. In too many accounts, the machine drops from the sky and has an impact. In other analyses, only costs count, and rising labor costs or other factors induce new technology, as if incentives have a magical effect. If that were true, we could count on escaping fossil fuels the moment they become too expensive. Yet history demonstrates, again and again, technology invented before being adopted, or lagging centuries behind its perceived need. Induced innovation is a lopsided view that oversimplifies a complex process that is cultural and institutional, as well as economic. Economics of course play a role in technological change, but supply-and-demand relationships exist within larger structures that are themselves shaped by law and custom—and by available technology, which of course influences the costs of factors of production.[40]

Technology is an element of economic structure. It sets the limits of the possible. It also results from structure, from human choices that rely on not only economic rationality but also cultural assumptions unrecognized as such, institutions so ingrained that they seem natural. Methods operate as constraints, especially when confused with nature. The "way things are done" looks obdurate, unchanging, and essential. The history of tobacco technology demonstrates how the human-made environment shaped the way things are done. Bright Tobacco represents an example of this process, which historians of technology know exist all around us. Myths such as the ones surrounding Bright Tobacco require bursting if we are to see clearly the world around us and stop assigning into the category of natural, outside our control, those elements of objective reality that have human histories. The political economies of the colonial, antebellum, and postbellum periods created systems of production sturdy enough to serve as the basis for twentieth-century regulation. The methods looked natural, although in fact they had changed several times, in different periods, in response to shifting market relations and regulations, supply-and-demand relationships, and the institutions that influenced both sides of that transaction. After all, technology is the principal means by which culture enters into the large economic structures assumed to control everything else. In technological choices, base and superstructure intertwine. That is why technology is so important, and yet so rarely understood.

The Real Thing

Tobacco Genetics

Race: a variety of such fixity as to be reproduced from seed.
Benjamin Daydon Jackson, A Glossary of Botanic Terms, with
Their Derivation and Accent

"Many botanists doubt the existence of plant species," began a letter published in 2006 in *Nature*. Its authors had surveyed the literature, and they disagreed with those scientists who argued that species were "arbitrary constructs of the human mind, as opposed to discrete, objective entities that represent reproductively independent lineages or 'units of evolution.'" When plants of one species looked unlike each other, according to this rejoinder, the reasons were usually "polyploidy, asexual reproduction, and over-differentiation by taxonomists" rather than "contemporary hybridization." Statistical analysis of "the plant biosystemic literature" found that 70 percent of "taxonomic species" and 75 percent of "phenotypic clusters in plants" corresponded to "reproductively independent lineages." That settled the question: plant species exist, their members have not reproduced outside the species (just as dogs and cats do not together produce offspring), and these species often correspond to distinctions in visible characteristics, although not always. The letter employed several scientific disciplines. Sometimes, however, these disciplines reached different conclusions. For this reason, the debates still rage: the more precise our knowledge becomes, it sometimes seems, the more questions it uncovers. This appendix therefore sorts out the knowledge of tobacco currently achieved in the sciences and examines the reasons for the differences among the different disciplines.[1]

Species are one thing; varieties are quite another. While species represent "reproductively independent lineages," subspecies designations trouble many sciences. In some plants, of course, varietals really do code for distinct characteristics (characters, rather, to use the official terminology) that can be ascribed to seed—to the plant's genetic makeup, its DNA. Grapes are an excellent example: no one would mistake a syrah for a pinot gris. Even in this case, however, the long history of wine and grapes and the coevolution of the two tell a story not unlike that of tobacco. Economic supply-and-demand relationships developed over centuries; so did their regulatory environment, whether in the form of the nineteenth-century Appellation d'Origine Contrôlée of French wines or the less explicit colonial, state, federal, social, and cultural institutions of tobacco in the United States. Because some tobacco types have long histories in trade that do not match perfectly with their genetics, they are better called cultivars than varieties—a term scientists use that indicates man-made articles. Different sciences have different trajectories, however. They sometimes end up with different conclusions. These help account for controversies such as that in the 2006 letter to *Nature*.

Scientific knowledge has a history, too—or rather, several histories, of multiple disciplines. The 2006 letter to *Nature* drew on genetics, population statistics, and botanic taxonomy. It also relied on evolutionary theory and historical narrative, as when the letter writers claimed that "reproductively independent lineages" from the past have defined today's species. This appendix follows a similar path to capturing "the real thing." It splits science into several disciplines, each with its own history, each with its own methods for producing natural knowledge. The historical investigation of tobacco types lies in the text of this book. Now genetic, statistical, and taxonomical approaches each get a turn. Since many readers likely believe genetics to be the answer, genetics comes first—both empirical investigations and the most current explanatory theories. Distinctions among types come from differences among seeds, according to conventional wisdom, no matter the level of classification. Race, variety, species, and even genus, family, and order are the categories employed in taxonomy, but many readers likely think that such taxonomical classifications indicate inherited (and heritable) characteristics. In this view, taxonomy relies on genetics. Moreover, many readers likely reflexively consider genetics the gold standard among the scientific approaches to grasping nature. So we begin there.[2]

THE CURRENT STATE OF GENETIC KNOWLEDGE

This book's introduction quoted a gene sequencing paper that found "a low level of genetic diversity within and among cultivated tobacco types." Genetic differences do exist between the varietal types, however, and the "high level of genetic identity (>0.77) between the different types" is not so small a difference, given that humans and chimpanzees have some 99 percent of genes in common. The economic importance of tobacco in the early twentieth century, at the dawn of genetics as a science, and the simplicity of its genetic structure—its amphidiploidy (to be explained in a moment)— meant that it provided some experimental material to early geneticists. It remains the subject of many classroom introductions to plant breeding. Scientists are still using the plant to test the latest methods, and they still seem surprised by their findings. In the 2007 paper, the geneticists were examining a method called "intersimple sequence repeat and inter-retrotranspon amplification." They accessed tobacco plants and classified their accessions by "manufacturing quality traits." Their flue-cured and sun/air-cured samples came from China and the United States. Of eleven burley samples, all but one (from Zimbabwe) had grown in the United States. Although the method being tested did find distinctions between the categories of manufacturing qualities, these distinctions did not correspond much with "phenotypic variations" and "the DNA fingerprinting suggests that there is little variation."[3]

Current genetic theory adds clarity to the similarity among tobacco types. Humans are what twenty-first-century geneticists call diploid: possessing two sets of chromosomes, one from each parent. Plants, on the other hand, are very often polyploid—they contain more than two sets of chromosomes. According to one biochemist, "well over half of all species of flowering plants possess more than two sets of chromosomes," and the proportion rises to 78 percent when only crop plants—those grown for human consumption—are considered. Indeed, polyploidy has proven a crucial device in geneticists' use of evolution, particularly the history of domestication of plants for human use. Polyploidy seems linked to larger plants, with bigger fruits and grains, which made them attractive to humans practicing selection. Even some simple animals are polyploids, and "even humans are descended from ancient polyploids that subsequently adopted a pseudodiploid genome organization." Moreover, "genome doubling" has occurred often in the evolution of "all complex animals"; perhaps twice in human evolution, for example. These multiplications of genetic material maybe helped repair DNA when necessary, or created opportunities for "cross talk" among the "parental genomes" to establish the characteristics of individual organisms.[4]

Polyploidy provides genetic opportunities for greater variation among offspring, which may have contributed to evolution. Polyploidy explains how, among all those red-flowering plants in a single patch, one occasionally produces pink blooms instead. However, *Nicotiana tabacum* lacks the polyploidy that provides so much variation in the offspring of other plants. Fortunately, polyploidy is not the only cause of generational permutations. Outright mutations do sometimes appear; these were assumed responsible for the appearance of white burley in Civil War–era Kentucky, for example. In addition to chromosomes, a great deal of the physical material that we call DNA remains poorly understood—unassigned to physical processes. Some scientists suspect that it regulates the expression or transcription of those genes and combinations of genes more directly linked to observable traits. Genes code for plant characters, but when those characteristics appear, and what mechanism accounts for their emergence, is a process still undergoing theoretical construction by scientists. *N. tabacum* was a plant much used in early genetic analysis, and its tendency to "break the type" when moved to new locations contributed to the emerging science of heredity. Today, twenty-first-century geneticists are still examining the whole body of genetic material, including the still-mysterious processes of expressing characteristics: what turns them on and off.[5]

Nicotiana tabacum—the species containing all the U.S.-grown commercial tobacco types—is "a so-called amphidiploid; meaning, in it are combined the chromosomes of two species which have hybridized in the past." Each of the two species that hybridized to compose *N. tabacum* provided a set of chromosomes that then doubled, creating an allotetraploid—an organism having four sets of chromosomes, in two identical pairs—that functions as a diploid because of the two (doubled) sets of chromosomes. *Tabacum*'s twenty-four pairs of chromosomes are really twelve unique pairs, doubled. Moreover, each tobacco plant pollinates itself: "Pollen is not windborne and any cross fertilization occurring among plants is insect mediating." The "flower structure and size of the pollen" (with both male and female parts contained within the flower) "make this a self-pollinated plant." These reproductive mechanisms, together with human cultivation practices, have over time produced remarkably stable plant populations. Each plant basically clones itself (with minimal protections from nearby plants' parts), but deliberate crossing is easy and produces predictable results. This stability contributes to tobacco's value as a classroom demonstration, and as experimental material.[6]

A BRIEF HISTORY OF GENETICS

To historians of genetics, the twentieth century was the Century of the Gene, beginning with the famous rediscovery of Mendelism, but now turning away from genes and toward the expression and transcription of the characters those genes can produce. Yet twentieth-century genetics possessed (naturally) characteristics inherited from its tributary disciplines and fields of investigation. Practical animal breeding and plant hybridization antedated the twentieth-century elaboration of the Mendelian and Darwinian theories and advanced alongside the science. From Mendel and Darwin in the nineteenth century through August Weismann's germplasm to Wilhelm Johannsen's invention of the word "gene" in the first decade of the twentieth century, the emergent field moved from a view of heredity as a flow in which evolution kept on happening, and changes in genetic material occurred over the generations, to an idea of fixity of the hereditary unit, now named the gene. Along the way, important influences on the young field were forgotten. Agricultural breeders had transformed themselves into formal geneticists and established departments of genetics in universities: these shifts account for the success of the science. Such practical work established the pattern by which the breeding of future generations shaped scientists' interpretations of the past. Thomas Hunt Morgan and his students relied on fruit flies, the statistical analysis of their breeding results, and their specific characters for the earliest findings of the field.[7]

Therefore, early genetics rapidly entered the realm of population statistics and relied on breeding—the analysis of breeding results—for its descriptions and understandings of traits and their workings. Historians of science call the technologies of scientific discovery "instrumentation," and the use of breeding results in describing genetic structures meant that the peculiarities of living instruments shaped the scientific knowledge that early geneticists produced. Mutations begat the recognition of other mutations, and the evolutionary concerns ("vitality, fecundity, sex ratio, and growth rates") that had motivated the use of *Drosophila* flies in genetic experimentation rapidly disappeared in favor of more easily identifiable characters such as eye color, wing formation, and other morphological traits. Producing a laboratory supply of flies in itself created a "flood of mutants" with which the fly group contended and classified, turning them to "neo-Mendelian experimental heredity" rather than the concerns of evolutionary biology. Research into tobacco, likewise, took a similar form. Scientists used the results of breeding to demonstrate parental genetics, although the theoretical constructions for such conclusions had a tendency to lag behind practical needs.[8]

Investigations into tobacco genetics at first assumed that the genetics of varietal types matched up with economic uses, but they often found that introducing far-flung varieties to new locations led to "breaking the type," or "a progressive, cumulative effect of environment which represents a more or less gradual readjustment to changed environmental conditions." This explains how Kentucky seed brought to Louisiana could become Perique, as Killebrew had described in the 1880 census. At first, the geneticists thought that this "violent breaking up of type" indicated "reversions to earlier and unimproved types of tobacco." A few years later, the federal scientists would argue that "there is no true line of demarcation between inherited characters and fluctuations with respect to many characters of commercial significance." Some leaf would not burn, which of course shocked the scientists. Other notable problems were variability in leaf number and branching or suckering growth patterns. The appearance of these oddities in the stock of introduced varieties not only confused the investigators but also complicated cultivation and harvesting processes that relied on similar plants in the field.[9]

Efforts to cross better plants also confused the picture. In 1915, investigators bred two tobacco species to shed light on the mechanisms controlling flower size, only to discover that the concatenating effects of cultivation practices shaped the outcome. When the flower appeared (in the life cycle of the plant) contributed to its size, as did its environment—greenhouse or field, fertilized or not.[10] A few years later, the authors undertook the direct scrutiny of varietal distinctions. "The authors also found themselves in need of some definite information as to the Mendelian details involved in character differences within the species." They crossed varietals with distinct characters, sometimes carrying a cross out ten generations to see what would breed true and what would not. The results:

A general result of these investigations has been a demonstration of the complexity of difference from a genetic standpoint between any two of these so-called fundamental varieties of Tabacum. . . . The so-called fundamental varieties of Tabacum intercross freely and produce fully fertile progenies. They cannot genetically, therefore, be regarded as representing anything but a few very distinct genotypes. A demonstration that a few such varieties may contain within them the possibility by means of recombination of producing a host of secondary varieties does not really demonstrate that they are fundamental.

In fact when we consider the fugitive nature of our Tabacum varieties, except in so far as they are kept isolated by natural or artificial means, the conclusion appears inevitable that we must regard all our varieties as fundamentally equivalent from a genetic standpoint. The really significant problem in considering the species is the determination of how these allelomorphic contrasts have come into existence. These investigations throw no light upon that problem.[11]

The plant's genome explains the similarity across varietals, then. It does not explain the differences between the types that developed in historical circumstances, rather than laboratory investigations.

POPULATION STATISTICS, TAXONOMY, AND THE HISTORICAL EVIDENCE OF PLANT DISTRIBUTION

This study of tobacco genetics built on others, attempts to study plant distributions and probe the origins of species. While some early twentieth-century studies posited tobacco's Asian origins, a general consensus has by the twenty-first century agreed upon a plant that originated in some area in Central or South America. The distribution of tobacco on the American continents indicates that the Native Americans of the Chesapeake employed *Nicotiana rustica*—a species distinct from the *Nicotiana tabacum* that English settlers turned into their commercial staple for export. The legend of John Rolfe may in fact be true—although what he brought north to his fellows from the West Indies was, according to this population-genetics evidence, a different *species* rather than a different *variety*. *N. rustica* may have been prevalent only on the east coast of North America, however. Further inland, in the central Great Plains, the Pawnees practiced production of seed capsules from a species known as *N. quadrivalvis*. "Indian farmers to the east, who raised Nicotiana rustica, preferred the Plains' variety; and . . . tried to trade with the Plains tribes for the seeds, [but] the Plains farmers jealously kept the tobacco variety to themselves." Casual usage often calls these distinctions varieties. Before the breeding for disease resistance that has taken place since the New Deal, however, real distinctions occurred mostly at the species level.[12]

The names of these plants confirm that they are different species, rather than varieties. The binomial (or, among botanists, binary) nomenclature—the two-name taxonomical system that still identifies flora and fauna—uses the genus (*Nicotiana*) followed by the species (*tabacum*, *quadrivalvis*, or *rustica*). Moreover, taxonomy predates genetics by centuries. It originated in the Columbian Exchange, that shuffling of plants and animals across the seams of the globe, that result of the 1492 voyage of Christopher Columbus. Tobacco participated in that moment of contact. At first landfall in the New World, Columbus's sailors remarked on the smoking habits of the "Indians" they thought they were meeting. When Columbus's patrons realized that the land sighted was not India, they collected samples of their discoveries. Specimens plucked from the New World's newness kept European natural philosophers "happily drunk on data" for centuries. The systems of classification that naturalists had inherited from antiquity groaned under the strain, and modern botanical hierarchies appeared in order to keep up. Carl von Linné, from Uppsala, Sweden, called Linnaeus in the Latin of Enlightenment scholarship, became the grandfather of modern botanic taxonomy when he devised the system of binomial nomenclature still used today.[13]

In 1737, Linnaeus published *Critica Botanica*—his fifth book in two years—which undertook methodical criticism of ancient and enduring botanic naming practices. While his masterwork of taxonomic definition, the *Species Plantarum*, would not appear until 1753, the earlier text did important preparatory work by pruning away old and redundant information. Linnaeus worked to establish rules for distinguishing species from one another, mocking along the way his colleagues who perceived differences among species based on specious taxonomies. *Critica Botanica* contains a delightful passage criticizing size as a distinguishing characteristic because it is "misleading, vague, indefinite . . . there are no limits or boundaries—except as marking gradation: grades only, no fixed limits! all Size is relative, and . . . it is not scientific to make one plant a standard for one's study of another." The size of leaves should be no distinction either; similarly locality makes no difference, nor season, nor color, nor scent, nor taste, nor use. Within this diatribe it is possible to see not only the enfant terrible berating his brethren but also the scientist resisting the temptation to make his view of the natural world reflect human needs. Tobacco's taxonomy contradicts the scientific goals and methods of botanic classification.[14]

Despite the Enlightenment scientist's view of nature as essentially unchanging over time, the influence of historical experience crept into even the Linnaean classification. Although the Swede forbade the use of languages other than Latin for names of genera and species, old Greek usages occasionally persisted—although the saints' names assigned by religious scholastics did disappear. *Nicotiana* became the generic name for tobacco because Linnaeus named the plant for the man who first brought it to Europe. Nicot apparently did introduce tobacco into France, from Spain, in 1559 or 1560. The generic naming of an Enlightenment botanist, the starting point for what we know, replaced with the name of a European diplomat the "barbarous name . . . Petum" derived from indigenous knowledge that was suspect to the European conquerors. Old barbarous names disappeared. Thus, the work of Enlightenment rolled on, toward ever greater detail in the classification of nature according to brittle schemes.[15]

Nowadays taxonomists provide some of the deepest challenge to the geneticists and population statisticians who authored the 2006 letter to *Nature*. As one textbook in plant taxonomy explained it, even when species can be distinguished in terms of reproductively independent lineages, the groupings above the genus and below the species level have long occupied "a world of much greater uncertainty." Above the species level, plants fall into a genus, every genus is part of a family, each family part of an order. In the nineteenth century, the *Nicotiana* genus was put sometimes in the order Monogynia and sometimes (as is the practice today) in the order Solanaceae. Below the species level occur categories now understood as subspecies, variety, and form. Practitioners sometimes use the term "strain" to describe a genetic difference of subspecies (and maybe even subvarietal) designation, while others nowadays prefer "cultivar" to describe the human-made types that rely on cultivation by humans for their characteristics.[16]

Other terms, too, have disappeared: botanists speak only gingerly about race. It once occupied a useful position, as it distinguished a "variety of such fixity as to be reproduced from seed" from other types, without reproductive fixity. Varieties were simply "a sort or modification subordinate to species," with no assumptions made about the causes of that modification, or its heritability. As Linnaeus understood in the eighteenth century, and as the geneticists investigating flower size confirmed in 1915, environment shapes plant development.

To the plant, of course, human technology is part of the environment. The term for characters that reproduce over generations was not variety, but race. Notably the term "race" was "used also in a loose sense for related individuals without regard to rank," that is, without regard to whether those individuals were related at the level of species, genus, family, or subspecies, variety, or strain. Some diseases of tobacco plants have been classified into races, or rather, pathogens' ability to infect a cultivar carrying specific resistance genes have allowed scientists to group them into races. This is not intended to advocate a return to the racial classifications that flourished in the nineteenth and twentieth centuries, but to recognize that an abandoned taxonomical category at one time grasped how varietal designations came as much from human intercession as from seed.[17]

Science makes questions simple: do identifiable characteristics reproduce in predictable ways? The answer is both yes and no. The plant population is stable, but not for the obvious reasons. Genetics and flower structure contribute to tobacco's stability but fail to explain the different types of tobacco in any meaningful way. Anyone who has worked in the fields knows that desirable characteristics do not take shape on their own. The human effort that is necessary to move plants toward the production of desirable commodities deserves attention. It too has a history, one episode of which has been recounted in the text of this book. Varietal designations are the products of human effort and human choices—technologies—which not only define the tobacco types but also shape our perception of nature.

Notes

ABBREVIATIONS

AHR	*American Historical Review*
BHR	*Business History Review*
BRFAL	Bureau of Refugees, Freedmen, and Abandoned Lands Records, National Archives and Records Administration, Microfilm Publications
ECU	East Carolina University, Special Collections, Greenville, North Carolina
FHS	Filson Historical Society, Louisville, Kentucky
GPO	U.S. Government Printing Office
HBS	The Baker Library, Harvard Business School, Cambridge, Massachusetts
JAH	*Journal of American History*
JEH	*Journal of Economic History*
JNH	*Journal of Negro History*
JSH	*Journal of Southern History*
KYLM	The Kentucky Library and Museum, Western Kentucky University Libraries, Bowling Green, Kentucky
LOC	Library of Congress, Chronicling America Collection, http://chroniclingamerica.loc.gov/
LOV	Library of Virginia, Richmond, Virginia
NARA DC	National Archives and Records Administration, Washington, D.C.
NARA MD	National Archives and Records Administration, College Park, Maryland
NCC	North Carolina Collection, UNC-Chapel Hill, Chapel Hill, North Carolina
NCHR	*North Carolina Historical Review*
NCSU	Special Collections, North Carolina State University, Raleigh, North Carolina
PLSC	Perkins Library Special Collections, Duke University, Durham, North Carolina
QJE	*Quarterly Journal of Economics*
SHC	Southern Historical Collection, UNC-Chapel Hill, Chapel Hill, North Carolina
USC	South Caroliniana Library, University of South Carolina, Columbia, South Carolina
VHS	Virginia Historical Society, Richmond, Virginia
VMHB	*Virginia Magazine of History and Biography*

INTRODUCTION

1. Ian Gately, *Tobacco* (Grove Press, 2001); G. Cabrera Infante, *Holy Smoke* (Harper & Row, 1985); David T. Courtwright, *Forces of Habit* (Harvard University Press, 2001); Jan Rogoziński, *Smokeless Tobacco in the Western World, 1550–1950* (Praeger, 1990); Allan M. Brandt, *The Cigarette Century* (Basic Books, 2007); Jarrett Rudy, *The Freedom to Smoke* (McGill-Queen's University Press, 2005); Cassandra Tate, *Cigarette Wars* (Oxford University Press, 1999).

2. Howard Cox, *The Global Cigarette* (Oxford University Press, 2000); Barbara Hahn, "Paradox of Precision," *Agricultural History* 82 (Spring 2008); Nannie May Tilley, *The Bright-Tobacco Industry* (University of North Carolina Press, 1948), 12, 130–54.

3. Pete Daniel, *Breaking the Land* (University of Illinois Press), 23; www.bae.uky.edu/Ext/Tobacco/PDFs/BurlMech.pdf (accessed Sept. 20, 2010).

4. Thomas P. Hughes, *Networks of Power* (Johns Hopkins University Press, 1983); Wiebe E. Bijker, Thomas P. Hughes, and Trevor J. Pinch, *The Social Construction of Technological Systems* (MIT Press, 1987); Bruno Latour, *Reassembling the Social* (Oxford University Press, 2005).

5. Nicholas Wade, "A Decade Later, Genetic Map Yields Few New Cures," *New York Times*, June 12, 2010.

6. W. W. Garner, H. A. Allard, and E. E. Clayton, "Superior Germ Plasm in Tobacco," in USDA *Yearbook* 1936 (GPO, 1937), 819; B. C. Yang et al., "Assessing the Genetic Diversity of Tobacco Germplasm," *Annals of Applied Biology* 150 (published online June 5, 2007), 393.

7. Charles E. Gage, "American Tobacco Types, Uses, and Markets," USDA Circular No. 249 (Jan. 1933; rev. ed. GPO, 1942).

8. "Type Classification of American-Grown Tobacco," USDA Miscellaneous Circular No. 55 (GPO, 1925).

9. Beth Cannady, "We Own Our Land," 248/14426, oral history interview, Folder 8, Leonard Rapport Papers, SHC; C. C. Davis to R. A. Green, July 19, 1933, Folder Apr. 1–July 31, 1933; Subject Correspondence Files, 1933–1935; Entry 2, RG 145; Agricultural Adjustment Administration Production Control, Tobacco, NARA MD. Hughes's *Networks of Power* contains the technological statement of systems theory; on Big Business, see Alfred D. Chandler, Jr., *The Visible Hand* (Harvard University Press, 1977).

10. Martin Hill & Co to Wm. H. Burwell, Oct. 25, 1879, Folder 49, Box 2, Series 1.2, Burwell Family Papers, SHC; J. H. Owen to William T. Sutherlin, Sept. 20, 1861, Folder 13, Box 2, William Thomas Sutherlin Papers, SHC; "For Tobacco Growers! Great Inducements Offered in Virginia, near Petersburg," *Progressive Farmer*, Dec. 2, Dec. 9, Dec. 16, and Dec. 23, 1890; Jan. 27, 1891; always on p. 5; J. B. Killebrew and Herbert Myrick, *Tobacco Leaf* (Orange Judd, 1897; repr. 1934), 339. Killebrew's expertise can be judged by his authorship of the 1880 census article on tobacco agriculture.

11. Thomas W. Crowder to William Gray, Oct. 18, 1846, William Gray Papers, VHS.

PROLOGUE

1. Nannie May Tilley, *The Bright-Tobacco Industry* (University of North Carolina Press, 1948), 24; J. B. Killebrew, "Report on the Culture and Curing of Tobacco in the United States," in *Report on the Productions of Agriculture as Returned at the Tenth Census*, vol. 3 (GPO, 1883), 704; M. Ruth Little, An Inventory of Historic Architecture, Caswell County, North Carolina (North Carolina Department of Cultural Resources, 1979), 162; Eldred E. Prince, Jr., in his *Long Green* (University of Georgia Press, 2000), 47, uses a South Carolina Department of Agriculture report as his source.

2. Angela Lakwete, *Inventing the Cotton Gin* (Johns Hopkins University Press, 2005); Thomas P. Hughes, *Networks of Power* (Johns Hopkins University Press, 1983). For dates, see the sources cited above.

3. John K. (Jack) Brown, "Louis C. Hunter's *Steamboats on the Western Rivers*," review, *Technology and Culture* 44 (Oct. 2003), 787; Carroll W. Pursell, ed., *Technology in America*, 2nd ed. (MIT Press, 1990).

CHAPTER 1: MAKING TOBACCO VIRGINIAN

1. John J. McCusker and Russell R. Menard, *The Economy of British America, 1607-1789* (University of North Carolina Press, 1985; repr. 1991), 118; the first tobacco warehouse in Pittsylvania County has been dated to 1793; see E. M. Sanchez-Saavedra, "A Preliminary Checklist of Tobacco Inspections 1680-1820," *Quarterly Bulletin, Archeological Society of Virginia* 24 (Sept. 1969): 80.

2. Robert Wilson, "Pittsylvania County 1785-," Account Book S-8, s.v. Capt. Charles Williams, June 9, 1785, p. 46, Robert Wilson Account Books, SHC; Jacob M. Price and Paul G. E. Clemens, "A Revolution of Scale in Overseas Trade," *JEH* 47 (Mar. 1987): 15, 17; Russell R. Menard, "The Tobacco Industry in the Chesapeake Colonies, 1617-1730," *Research in Economic History* 5 (1980): 109-77.

3. Those who draw firm distinctions between economic and cultural causes include David Eltis, Philip Morgan, and David Richardson, "Agency and Diaspora in Atlantic History," *AHR* 112 (Dec. 2007): 1329-58.

4. Lois Green Carr and Russell R. Menard, "Land, Labor, and Economies of Scale in Early Maryland," *JEH* 49 (June 1989): 413; John M. Staudenmaier, S.J., *Technology's Storytellers* (MIT Press, 1985), introduction and chap. 1.

5. Lewis Cecil Gray, *History of Agriculture in the Southern United States to 1860* (Carnegie Institution, 1933), 1:21-22, 1:213-33; Avery O. Craven, *Soil Exhaustion as a Factor in the Agricultural History of Virginia and Maryland, 1606-1860* (University of Illinois Press, 1925). Viewing techniques in terms of work processes follows the lead of Ruth Schwartz Cowan, *More Work for Mother* (Basic Books, 1985).

6. Gray, *History of Agriculture*, 1:215-17; Thomas Singleton, *A Treatise on the Culture of Tobacco* ([n.p.], 1770?), 15-16, 19; William Tatham, *An Historical and Practical Essay on the Culture and Commerce of Tobacco* (Vernor and Hood, 1800; repr. University of Miami Press, 1969), 8, quotations at 13; James Jennings, *A Practical Treatise on the History, Medical Prop-*

erties, and Cultivation of Tobacco (Sherwood, Gilbert, and Piper, 1830), 39, 84, 87, 92–93; Thomas Glover, *An Account of Virginia* (Royal Society, 1676), 28–29.

7. Tatham, *Culture and Commerce*, 9, 13; E. Pendleton to John Ambler, Apr. 15, 1826, Edmund Pendleton Papers, VHS.

8. Jack P. Greene, ed., *Diary of Colonel Landon Carter* (VHS and University Press of Virginia, 1965), 1:140, s.v. Feb. 9, 1757; Peter Coclanis, "Thickening Description," *Agricultural History* 64 (Summer 1990): 9–16.

9. Tatham, *Culture and Commerce*, 14–17.

10. Ibid., 14, 16; Singleton, *Treatise*, 20.

11. Greene, *Diary of Landon Carter*, 1:149, s.v. Mar. 4, 1757; S. Max Edelson, *Plantation Enterprise in Colonial South Carolina* (Harvard University Press, 2006).

12. Gray, *History of Agriculture*, 1:217; Tatham, *Culture and Commerce*, 22; Singleton, *Treatise*, 20; Greene, *Diary of Landon Carter*, 1:158–59, s.vv. Apr. 2, 1757, June 8, 1757.

13. Singleton, *Treatise*, 21; Peter McEnery, "Tobacco Factory Daybook," Dec. 1837–Jan. 1857, pp. 66, 190, s.vv. Oct. 18, 1838; Oct. 3, 1839, Thomas Branch & Co. Papers, VHS.

14. Greene, *Diary of Landon Carter*, 1:162, s.vv. July 14, 1757, July 18, 1757; Singleton, *Treatise*, 21–22.

15. Tatham, *Culture and Commerce*, 20; Singleton, *Treatise*, 11–12, 22.

16. Tatham, *Culture and Commerce*, 24; Singleton, *Treatise*, 23; Jennings, *Practical Treatise*, 39.

17. N. Atkinson to Sir [Berkeley], Sept. 23, 1810, William N. Berkeley Papers, PLSC; Greene, *Diary of Landon Carter*, 2:774, s.v. Sept. 19, 1773. *N. tabacum* is considered a perennial, but it is usually grown as an annual.

18. Tatham, *Culture and Commerce*, 23.

19. Robert Hamilton to Samuel Tarry, July 20, 1824, Tarry Family Papers, SHC; J. & D. Walker to Chris Winfree, Apr. 19, 1845, Folder 7, Box 1, William E. Uzzell Papers, SHC. As an example of how tricky moisture makes tobacco quantification, witness Price and Clemens, "Revolution of Scale," 23, which demonstrates little grasp of that "unfathomable" and "mysterious rubric 'order'" despite the author's erudition in the tobacco trade. See his "Merchants and Planters," in *Tobacco in Atlantic Trade* (Ashgate, 1995), for a correction of this misapprehension. This difficulty of measurement may also explain why the U.S. government declared of tobacco in 1842 that "we have to grope in the dark for statistical conclusions upon a subject affecting the interests of a million of people connected with the cultivation and manufacture of this important staple." Secretary of the Treasury, "Tobacco," May 28, 1842, H.doc. 235 (27-2) 404, p. 6.

20. A. Petricolas & Co. to Messrs. Peters & Whitehead, Sept. 2, 1830, Folder 1, Box 1, Floyd L. Whitehead Papers, SHC; Singleton, *Treatise*, 24.

21. Gray, *History of Agriculture* [quoting Glover 1676], 1:22; 1:216–17; Singleton, *Treatise*, 24.

22. Singleton, *Treatise*, 25; Tatham, *Culture and Commerce*, 25.

23. Singleton, *Treatise*, 24–25; Jennings, *Practical Treatise*, 39; Tatham, *Culture and Commerce*, 29–34.

24. Tatham, *Culture and Commerce*, 36; Singleton, *Treatise*, 27; Joseph C. Robert, *The Tobacco Kingdom* (Duke University Press, 1938), 41.

25. Tatham, *Culture and Commerce*, 37–38, quotations at 38, 40.

26. Ibid., 37–41, 69–106, quotations at 38, 40, 41; Singleton, *Treatise*, viii; Robert, *Tobacco Kingdom*, 235–37; Raymond Jahn, *Tobacco Dictionary* (Philosophical Library, 1954), 130, s.v. prizing; "Memorandum of Tobacco prized 1848," Folder 2, Box 1, Floyd L. Whitehead Papers.

27. Frederick F. Siegel, *Roots of Southern Distinctiveness* (University of North Carolina Press, 1987), 62; Price and Clemens, "Revolution of Scale," 21. Carr and Menard's Chesapeake system (in "Land, Labor, and Economies of Scale") lasts a single calendar year, but there is little evidence that marketing work ended in December.

28. Robt. Murray to John Corlis, Jan. 30, 1788, Folder 3, Corlis-Respess Family Papers, FHS.

29. Virginia General Assembly, *Journals of the House of Burgesses* (Virginia State Library, 1905–1915), 1:17, 1:47, 1:53.

30. "A Dialogue between Thomas Sweet-Scented, William Oronoco … ," pamphlet, 3rd ed., 1732, p. 10; Allan Kulikoff, *Tobacco and Slaves* (University of North Carolina Press, 1986), 107–14, provides evidence of violent opposition to even the eighteenth-century version of the law. Charles E. Gage, "Tobacco, Tobacco Hogsheads, and Rolling Roads," paper read before the Falls Church Historical Commission, July 18, 1959, p. 4.

31. R. Douglas Hurt, *Indian Agriculture in America* (University Press of Kansas, 1987), 45–48, 61.

32. Gray, *History of Agriculture*, 1:69–73, quotation at 1:70; Jacob M. Price, *France and the Chesapeake* (University of Michigan Press, 1973), 1:314, 1:319–20; "The Cultivation of Tobacco," De Bow's *Review* 9 (Aug. 1855): 241–42.

33. Tatham, *Culture and Commerce*, 20; George J. F. Clarke, "On Spanish Tobacco," *Southern Agriculturalist* 1 (June 1828): 259; W. H. Simmons, "On the Cultivation of Cuba Tobacco, by Joseph M. Hernandez," *Southern Agriculturalist* 3 (Sept. 1830): 458–67; Jennings, *Practical Treatise*, 40.

34. Howard Bodenhorn, "A Troublesome Caste," *JEH* 59 (Dec. 1999): 978; Russell Menard, *Sweet Negotiations* (University of Virginia Press, 2006), xii, 30, quotation at 4; Carr and Menard, "Land, Labor, and Economies of Scale," 410; Ralph V. Anderson and Robert E. Gallman, "Slaves as Fixed Capital: Slave Labor and Southern Economic Development," *JAH* 64 (June 1977): 24–46.

35. This time line relies heavily on works by Jacob Price, as collected in *Tobacco in Atlantic Trade*. In "Merchants and Planters," 11, he avers that little evidence exists for which marketing system prevailed when. On the shakeout, too, see Price and Clemens, "Revolution of Scale," 6–7, 18–19. Gloria L. Main, *Tobacco Colony* (Princeton University Press, 1982); T. M. Devine, *Tobacco Lords* (John Donald, 1975); Arthur Pierce Middleton, *Tobacco Coast* (Mariners' Museum, 1953); T. H. Breen, *Tobacco Culture* (Princeton University Press, 1985).

36. Gary Biglaiser, "Middlemen as Experts," *RAND Journal of Economics* 24 (Summer 1993): 212–23, is part of the literature on how information affects transactions, derived from George A. Akerlof, "The Market for 'Lemons': Quality Uncertainty and the Market Mechanism," *QJE* 84 (Aug. 1970): 488–500.

37. Gray, *History of Agriculture*, 1:218; Tatham, *Culture and Commerce*, 4–5.

38. Breen, *Tobacco Culture*; McCusker and Menard, *Economy of British America*, 124; Jacob M. Price, *Capital and Credit in British Overseas Trade* (Harvard University Press, 1980), 96–97; Price, *France and the Chesapeake*, 1:509, 1:610–11, tables 6 (B) and 7 (B), 2:845–48.

39. Secretary of the Treasury, "Tobacco," 19, 48; according to my calculations on the data in this source, the average annual figures between 1824 and 1833 were 19,077,716 pounds imported for Britons' internal consumption and 14,238,551 pounds imported for re-exportation elsewhere. Imperial European nations established botanic gardens to collect specimens of commodity plants for selection, breeding, and transfer of strains with desirable characteristics. See Daniel R. Headrick, *Tentacles of Progress* (Oxford University Press, 1988), 211–15, 219–29. Tatham, *Culture and Commerce*, 1, 4; Jennings, *Practical Treatise*, mentions Ireland on iii–iv. On tariffs, stems, and smuggling, see Robert B. Campbell to R. M. T. Hunter, Mar. 16, 1855, written on Horatio N. Davis, "Tobacco," price-circular, Mar. 1, 1855, Correspondence, Robert Mercer Taliaferro Hunter Papers, VHS.

40. J. & D. Walker to Chris Winfree, Apr. 19, 1845, Folder 7, Box 1, William E. Uzzell Papers; Thomas W. Crowder to William Gray, Oct. 18, 1846, William Gray Papers, VHS; on curing methods, see Robert, *Tobacco Kingdom*, 38–46.

41. Singleton, *Treatise*, vi–vii; Gray, *History of Agriculture*, 2:762–63; Price, *France and the Chesapeake*.

42. Carolina's backcountry settlers, too, more easily floated produce down to Charleston. William K. Boyd, introduction to *William Byrd's Histories of the Dividing Line* (Dover, 1967), xxix–xxx; Lindley S. Butler and Alan D. Watson, eds., *The North Carolina Experience* (University of North Carolina Press, 1984), 83; Gray, *History of Agriculture*, 1:43; receipt for leaf sent from Plymouth (on the Albemarle Sound) to Richmond, 1819, Brownrigg Family Papers, ECU.

43. Gray, *History of Agriculture*, 1:221; Virginia, *Journals of the House of Burgesses*, 2:23, 2:29, 2:38, 2:145; "Waaren-Preise," price-circular, Copenhagen, Nov. 17, 1804, Apr. 30, 1805, Tompkins Family Papers, VHS.

44. Dennistown Hill & Co. to John Wesley Hunt, June 8, 1846, John Wesley Hunt Papers, FHS; "The Tobacco Trade," De Bow's *Review* 2 (July 1846): 42–46; "New Tobacco Inspection Laws," *DeBow's Review* 2, no. 5 (Nov. 1846): 355–60. On Spanish hunger for the goods of Kentucky settlers, see materials in the Pontalba Papers, Wilkinson-Miro Correspondence; Wilkinson's Memorial in the Temple Bodley Collection; all, translated and transcribed by Gilbert Pemberton in the 1910s from the Louisiana Historical Society, New Orleans, now lodged in the FHS. The originals were destroyed in France during the First World War. Many thanks to Kathleen DuVal for providing me with these materials.

45. Gray, *History of Agriculture*, 1:218, quotation at 1:236; Alexr. B. Barret to David Bullock Harris, Feb. 10, 1843, Folder Correspondence/Papers 1843, Jan.–May, David Bullock Harris Papers, PLSC.

46. "Waaren-Preise"; Robert B. Campbell to R. M. T. Hunter, Mar. 16, 1855, Robert Mercer Taliaferro Hunter Papers.

47. Thomas Branch & Bro. to James Hunter & Co., Nov. 7, 1846, letterbook, May 10, 1846–Aug. 16, 1849, Thomas Branch & Co. Papers; Jones & Blair to Wm. Shanks, June 12, 1846, Folder 21, Box 2, Shanks Family Papers, SHC.

48. "Sales Eights Hogsheads Tobacco by Deane & Brown," account sales, Sept. 23, 1854, Allen Family Papers, VHS; Samuel Nowlin to Dickenson, Pannill & Co., May 27, 1828, Folder 2, Box 1, Robert Wilson Papers, SHC.

49. A. Petricolas & Co. to Messrs. Peters & Whitehead, Sept. 16, 1830, Folder 1, Box 1, Floyd L. Whitehead Papers; Thomas Branch & Bro. to Brooking Elder, May 29, 1846, letterbook, May 10, 1846–Aug. 16, 1849, Thomas Branch & Co. Papers.

CHAPTER 2: GROWING THE BUSINESS

1. Peter Wainwright & Jack Co., agreement with Charles Whittingham [or Worthington], Apr. 18, 1814; Chas. Whittingham to Peter Wainwright, Mar. 24, 1816; both, Folder 1767–1816, P. F. Wainwright Papers, PLSC.

2. Joseph C. Robert, *The Tobacco Kingdom* (Duke University Press, 1938), 174; Franklin F. Mendels, "Proto-Industrialization: The First Phase of the Industrial Process," *JEH* 32 (Mar. 1972): 241–61.

3. Secretary of the Treasury, "Tobacco," May 28, 1842, H.doc. 235 (27-2) 404, p. 5; Lewis Cecil Gray, *History of Agriculture in the Southern United States to 1860* (Carnegie Institution, 1933), 1:443.

4. Economic historians have long recognized the region's industrial past and its status as the sixth-most-industrialized region in the early nineteenth-century world; such analysis has not penetrated the larger scholarship. Robert William Fogel and Stanley L. Engerman, *Time on the Cross* (Little, Brown, 1974), 255. Claudia Goldin, *Urban Slavery in the American South, 1820–1860* (University of Chicago Press, 1976), should have buried the image of the nonindustrial South, yet still econometric scholars of the South's failure to industrialize sometimes claim that the tobacco industry "drew upon agricultural materials for inputs and the farm community for final market," an assumption for which the authors provided a deplorable scarcity of evidence. Fred Bateman and Thomas Weiss, *A Deplorable Scarcity* (University of North Carolina Press, 1981), 20. Few words or numbers illustrate manufactured tobacco returning to the farm or even the urban hinterland. Likewise, Jeremy Atack and Peter Passell, *A New Economic View of American History*, 2nd ed. (W. W. Norton, 1994), 355–56; and John Majewski, *A House Dividing* (Cambridge University Press, 2000), 147–48, ascribe the South's failure to industrialize mostly to the "slow growth of an indigenous mercantile community." Richard C. Wade, *Slavery in the Cities* (Oxford University Press, 1967); Suzanne Schnittman, "Slavery in Virginia's Urban Tobacco Industry, 1840–1860" (PhD diss., University of Rochester, 1987); Frank Towers, *The Urban South and the Coming of the Civil War* (University of Virginia Press, 2004).

5. John H. Garber, "Tobacco," in U.S. Bureau of the Census, *Twelfth Census of the United States*, vol. 9, *Manufactures*, pt. 3 (GPO, 1902), 637–72, quotation at 672. Most compilations of data on industrial output rely on the value added to raw materials by manufacturing. Since the processes of manufacturing cigars and tobacco were entirely different, however, such a method provides little useful information.

6. Henry O. & Peter McEnery to Brandon, McKenna & Lea, May 4, 1831, letterbook of Henry O. & Peter McEnery, Apr. 6, 1830–July 6, 1838, Peter McEnery Papers, VHS (hereafter McEnery letterbook 1830–1838). For the range of manufactured tobacco products, see

U.S. Secretary of State, *Fourth Census of the United States*, vol. 2 (GPO, 1823), 8–31; McEnery & McCulloch to Smiler & Sunk, June 3, 1852; McEnery & McCulloch to Chas. R. Somervail Co., Sept. 4, 1852; both, letterbook of McEnery & McCulloch, Dec. 5, 1851–May 30, 1853, Peter McEnery Papers, VHS (hereafter McEnery letterbook 1851–1853).

7. Henry Fitzhugh to James Russell, ca. Nov. 29, 1760 (letter not sent, crossed out); Henry Fitzhugh to James Buchanan, June 7, 1748; both, letterbook 1746–1774, Henry Fitzhugh Papers, PLSC. *Twelfth Census*, 3:815; Jack P. Greene, ed., *Diary of Colonel Landon Carter* (VHS and University Press of Virginia, 1965), 1:84, s.v. Mar. 13, 1752; Victor S. Clark, *History of Manufactures in the United States, 1607–1860* (Carnegie Institution, 1916), 18–19; Jacob M. Price, "The Beginnings of Tobacco Manufacture in Virginia," *VMHB* 64 (Jan. 1956): 3–29.

8. Land patents, deeds, and plats in Henry County for different branches of the family date from 1759; see Folders 665, 666, and 667, Box 20; estate inventories and wills can be found in Folders 668–72, 676, and 677, Box 20; and a "Record of Slave Births" and "Negroes births" dating from the 1820s is located in Folder 650, Box 20; all, Gravely Family Papers, Personal Papers Collection, LOV. The Hairstons have recently been the subject of a powerful popular history: Henry Wiencek, *The Hairstons* (St. Martin's Press, 1999).

9. P. & G. Gravely, "Tobacco Book," accounts sales, 1823–1836, Folder 645, Box 19; "Bill of Complaint and Answer in B. F. Gravely & Sons vs. B. F. Gravely & Co.," Folder 662, Box 20; Geo. Gravely to Peyton Gravely, Feb. 6, 1827, Folder 692, Box 21; "Facts Worth Knowing," pamphlet, 1881, Folder 662, Box 20; see also the 1876 stencils in Folder 481a, Box 20; all, Gravely Family Papers.

10. Joseph A. Martin, *A New and Comprehensive Gazeteer of Virginia and the District of Columbia* (Joseph Martin, 1835), 258, quoted in Nannie May Tilley, *The R. J. Reynolds Tobacco Company* (University of North Carolina Press, 1985), 5–7.

11. Frederick F. Siegel, *Roots of Southern Distinctiveness* (University of North Carolina Press, 1987), 27; Collection Description; W. T. Sutherlin to Stokes & Law, note for rent of tobacco factory, Sept. 21, 1849, Folder 1, Box 1; H. W. Holcombe, "Accounts," 1850, for manufacturing ingredients such as cochineal and gum Arabic, Folder 1, Box 1; all, William Thomas Sutherlin Papers; "Memorials of the Life, Public Services, and Character of William T. Sutherlin" (Dance Bros., 1894), 7–9, 34–35, PLSC.

12. R. G. Dun & Co., *Virginia*, 43:279, s.v. Allen & Preston, entry for June 3, 1857; 43:162, s.v. John F. Allen, entry for Feb. 24, 1853; 43:163, s.v. Ballanff & Co, entry for Jan. 15, 1858; R. G. Dun & Co. Collection, HBS.

13. "Description and Guide," 2, 3, 5, Thomas Branch & Co. Papers; Henry O. & Peter McEnery to Brandon, McKenna & Lea, May 4, 1831, McEnery letterbook 1830–1838; McEnery & McCulloch to Brandon Williams & Co., Dec. 22, 1851, McEnery letterbook 1851–1853.

14. For his dates, see the finding guide to his papers at PLSC. He was the tobacco manufacturer who figured so largely in Robert S. Starobin, *Industrial Slavery in the Old South* (Oxford University Press, 1970), which covered many sorts of industrial production and was part of the movement to understand the relationship between slavery and industrialization. Glenn Porter and Harold C. Livesay, *Merchants and Manufacturers* (Johns Hopkins University Press, 1971), quotation at 5n5, demonstrated the connections between the com-

mercial and industrial sectors. David Hancock, *Citizens of the World* (Cambridge University Press, 1995), 130–31, has recently (and correctly) assailed their careful taxonomy of types of merchants as too rigid to reflect reality, yet their sense of the multiple roles of market mediation performed by merchants still holds true.

15. Terence K. Hopkins and Immanuel Wallerstein, "Commodity Chains in the World-Economy prior to 1800," *Sociological Review* 10 (Summer 1986): 157–70; Gary Gereffi and Miguel Korzeniewicz, eds., *Commodity Chains and Global Capitalism* (Greenwood Press, 1994), 2, 34–47, 123–41; Alex Hughes and Suzanne Reimer, eds., *Geographies of Commodity Chains* (Routledge, 2004); Warren Belasco and Roger Horowitz, eds., *Food Chains* (University of Pennsylvania Press, 2008).

16. McEnery & McCulloch to Brandon Williams & Co., Apr. 13, 1852, McEnery letterbook 1851–1853.

17. "American Tobacco Trade," *DeBow's Review* 3 (June 1847): 575–79, quotation at 575.

18. Peter McEnery to Dr. Snead, July 20, 1835, McEnery letterbook 1830–1838.

19. Christian & Lea to W. T. Sutherlin, Jan. 30, 1862, Folder 15, Box 2, Sutherlin Papers; Robert L. De Coin, *History and Cultivation of Cotton and Tobacco* (Chapman and Hall, 1864), 292–96; Edward Pollock, *Sketchbook of Danville, Virginia* (1885, repr. Danville Historical Society, 1976), 120.

20. Turner & Duval to William Gray, Apr. 23, 1845, William Gray Papers, VHS; Henry O. & Peter McEnery to Thorburn & Smith, Aug. 2, 1830, McEnery to Brandon, McKenna & Lea, May 4, 1831, McEnery to Thorburn & Smith, Jan. 8, 1833; all three, McEnery letterbook 1830–1838; Edwin Tunis, *Colonial Craftsmen and the Beginnings of American Industry* (World Publishing Co., 1965), 54; "Snuff ready for sweating," in Nathaniel Bowe, "Diary," s.v. Oct. 27, 1846, SHC.

21. J. J. Heriges, "Making Plug Tobacco," U.S. Patent No. 6581X, granted June 13, 1831; Kennett, Dudley, & Co. to W. T. Sutherlin, Jan. 10, 1860, Folder 4, Box 1, Sutherlin Papers.

22. John T. Oxley, circular letter, May 1, 1841, Folder 16, Towles and Soaper Papers, FHS; the section "Imports, Deliveries, and Stock" lists "Virginia and Kentucky" as one category of hogshead, "Negro Head" as another, "Turkey" and "Florida" as others, in Horatio N. Davis, "Tobacco," price-circular, Mar. 1, 1855, Correspondence, Robert Mercer Taliaferro Hunter Papers, VHS.

23. Dating the use of brand names (and the arrival of a mass market for consumer goods) has proven tricky: Lizabeth Cohen, *A Consumer's Republic* (Knopf, 2003); Susan Strasser, *Satisfaction Guaranteed* (Pantheon Books, 1989), 18–19; T. H. Breen, *The Marketplace of Revolution* (Oxford University Press, 2004). Treatments of branding are largely unsatisfying, although they range from the hagiographical, such as David Powers Cleary, *Great American Brands* (Fairchild, 1981), to the profoundly theoretical, such as Michael R. Baye and Jon P. Nelson, eds., *Advertising and Differentiated Products* (JAI Press, 2001). On the other hand, Teresa da Silva Lopes, *Global Brands* (Cambridge University Press, 2007), succeeds in linking branding to business strategy. It is worth remarking that tobacco contains a drug, which people pursue in ways defined today as addictive. As a result, demand was assured, and branding probably only partly worked to associate certain qualities with a particular name or product.

24. Geo. W. Parrish, "Dr. William Pendleton, Bot of J. B. Parrish & Bros," receipt, Oct. 23, 1830, Folder Correspondence/Papers 1774–1845, Box 1, Madison and William James Pendleton Papers, 1775–1932, PLSC; Robert Wilson, "Pannill Wilson & Co. Blotter for 1834," 415, vol. 50, Folder 50, Robert Wilson Account Books, SHC; similar account books for country stores across the nation demonstrate the sale of tobacco as a generic good without a brand name at the consumer level.

25. Davis, "Tobacco," price-circular, Mar. 1, 1855; Jeffrey Ruggles, *The Unboxing of Henry Brown* (Library of Virginia, 2003), 11. For further examples of manufactured tobacco exported to distant markets, see T. & H. Messenger, "Tobacco Circular," New York price-circular, Feb. 2, 1859, General Collection, VHS; "Import of Manufactured Tobacco at New York," *DeBow's Review* 10 (May 1851): 547; on price-circulars and the importance of information, see John J. McCusker, "The Demise of Distance: The Business Press and the Origins of the Information Revolution in the Early Modern Atlantic World," *AHR* 110 (Apr. 2005): 295–319.

26. R. G. Dun & Co., *Virginia*, 43:279, s.v. Allen & Preston; Thomas Branch & Bro. to R. Y. Overly, Dec. 10, 1846, letterbook May 10, 1846–Aug. 16, 1849, Thomas Branch & Co. Papers.

27. Peter McEnery to Dr. Snead, Aug. 11, 1835; McEnery to Thomas Javier, Apr. 27, 1830; both, McEnery letterbook 1830–1838.

28. S. Wyatt & Co. to W. T. Sutherlin, Apr. 20, 1860, Folder 5, Box 1, Sutherlin Papers; McEnery & McCulloch to Adams & W. Chesney, Apr. 22, 1852; McEnery & McCulloch to Patterson & Price, Apr. 24, 1852, both, McEnery letterbook 1851–1853; Robert, *Tobacco Kingdom*, 217.

29. James D. Norris, *R. G. Dun & Co., 1841–1900* (Greenwood Press, 1978), 9–19. R. G. Dun was of course the predecessor to today's Dun & Bradstreet firm, engaged in similar work. Scott Sandage, *Born Losers* (Harvard University Press, 2005), 101, 109–12.

30. McEnery & McCulloch to Brandon Williams & Co., Apr. 13, 1852, McEnery letterbook 1851–1853.

31. Henry O. & Peter McEnery to Thorburn & Smith, Nov. 25, 1830, Dec. 31, 1830, McEnery letterbook 1830–1838; Jed. Lindsay to Sutherlin, Dec. 7, 1860, Folder 8, Box 1, Sutherlin Papers.

32. Archibald Williams Patterson, "Memoir, 1858–1933," VHS; Thomas Branch & Co., "Daybook 1837 Dec.–1857 Jan.," s.v. Sept. 8, 1838; s.vv. also especially June 2, 1838, Jan. 3, 1839 (for annual slave hire), Mar. 25, 1840, Commodity Books, Thomas Branch & Co. Papers.

33. Henry Box Brown, *Narrative of Henry Box Brown* (Boston, 1849; repr. Philadelphia: Rhistoric Publications, 1969), 36–42, quotations at 41, 42; Schnittman, "Slavery in Virginia's Urban Tobacco Industry," especially the appendices; Joseph C. Robert, "The Tobacco Industry in Ante-Bellum North Carolina," *NCHR* 15 (Apr. 1938): 125–26.

34. Sidney W. Mintz, *Sweetness and Power* (Viking Penguin, 1985); John A. Heitmann, *The Modernization of the Louisiana Sugar Industry, 1830–1910* (Louisiana State University Press, 1987); John C. Rodrigue, *Reconstruction in the Cane Fields* (Louisiana State University Press, 2001); Richard J. Follett, *The Sugar Masters* (Louisiana State University Press, 2005).

35. Peter McEnery to David Bridges, July 10, 1835, McEnery letterbook 1830–1838; Mary Wingfield Scott, *Old Richmond Neighborhoods* (William Byrd Press, 1950, repr. Valentine Museum, 1975).

36. W. T. Sutherlin, receipt for Mr. William Long's Crop of Tobacco, May 1849, Folder 1, Box 1, Sutherlin Papers; P. & G. Gravely, "Tobacco Book"; "Bill of Complaint"; Daniel F. Spulber, *Market Microstructure* (Cambridge University Press, 1999).

37. McEnery & McCulloch to Patterson & Price, Apr. 9, 1852, McEnery letterbook 1851–1853.

38. Columella, *On Agriculture*, 1:27, ranks credit as high in importance as willingness to work and knowledge of how to do so. The quote about credit supporting agriculture "as the rope supports the hanged" is sometimes ascribed to Louis XIV.

39. Scott, *Old Richmond Neighborhoods*, 82; Barbara Hahn, "Making Tobacco Bright" (PhD diss., University of North Carolina–Chapel Hill, 2006), chap. 2.

40. John P. Pleasants & Sons to Peyton Gravely, Feb. 13, 1838, and May 9, 1838, Folder 692, Box 21, Gravely Family Papers; A. Petricolas & Co. to Messrs. Peters & Whitehead, Sept. 16, 1830, Folder 1, Box 1, Floyd L. Whitehead Papers, SHC; Thomas Branch & Bro. to Brooking Elder, May 29, 1846; to James Hunter & Co., Nov. 7, 1846; both, letterbook May 10, 1846–Aug. 16, 1849, Thomas Branch & Co. Papers. The Petricolas letter cited here makes it clear that the tobacco described as "good dry Shipping Tobacco's & very sweet" and that "of a Manufacturing quality" are from the same crop, consecutive hogsheads from a single grower, while the May 29, 1846, Branch letter is more vague: "Common lugs and common leaf are very low, but shipping qualities and manfrs. are in good demand."

41. George Penn to Mary Penn, Jan. 24, 1851 or 1857, Folder 1850–1869, Green W. Penn Papers, PLSC.

42. Henry O. & Peter McEnery to Ashton & Moale, Apr. 23, 1830, McEnery letterbook 1830–1838; J. H. Owen [or Isham] to W. T. Sutherlin, Sept. 20, 1861, Folder 13, Box 2, Sutherlin Papers.

43. Henry Fitzhugh to Jno. & John B Mitchellson, May 8, 1747, May 27, 1748; both, letterbook 1746–1774, Henry Fitzhugh Papers; Read & Jeter to William Shanks, Mar. 9, 1860; N. M. Martin & Son & Co. to Wm. Shanks, Oct. [18], 1860; both, Folder 28, Box 3, Shanks Family Papers, SHC.

44. Finding Guide and Correspondence, David Bullock Harris Papers, PLSC.

45. Shipping Receipt, Apr. 26, 1841; C. Whittingham to D. B. Harris, Apr. 27, 1841; C. Whittingham to D. B. Harris, Apr. 25, 1841; A. B. Barret to D. B. Harris, Apr. 10, 1841; all, Folder Correspondence/Papers 1841 Jan.–May, Box 1, Harris Papers.

46. A. B. Barret to D. B. Harris, Feb. 10, 1843; Feb. 13, 1843; both, Folder Correspondence/Papers 1843, Jan.–May, Harris Papers.

47. David Bullock Harris to John Gilliat & Co., Aug. 8, 1847, Folder Correspondence/Papers 1847 July–Aug., Box 3, Harris Papers; Leland Smith, "A History of the Tobacco Industry in Kentucky from 1783 to 1860" (MA thesis, University of Kentucky, 1950), 30–31.

48. McEnery & McCulloch to Brandon Williams & Co., May 5, 1852; McEnery & McCulloch to Adams McChesney, May 5, 1852; both, McEnery letterbook 1851–1853.

49. McEnery & McCulloch to Brandon Williams & Co., May 5, 1852; McEnery & McCulloch to Adams McChesney, May 5, 1852; both, McEnery letterbook 1851–1853.

50. McEnery & McCulloch to Patterson & Price, Sept. 14, 1852; McEnery & McCulloch to Chas. R. Somervail Co., Sept. 4, 1852; both, McEnery letterbook 1851–1853.

51. McEnery & McCulloch to James Hunter & Co., Nov. 11, 1852, McEnery letterbook 1851–1853.

52. Coleman Wortham to W. T. Sutherlin, Mar. 3, 1862, Folder 16, Box 2, Sutherlin Papers.

53. Robert A. Martin to Wm. Henry Burwell, July 23, 1872, Folder 44, Box 2, Series 1.2, Burwell Family Papers, SHC.

54. Chas. D. De Ford & Co. to [Sutherlin], Feb. 15, 1860, Folder 4, Box 1, Sutherlin Papers.

CHAPTER 3: DEATH AND TAXES

1. Geo. W. Palmer to W. T. Sutherlin, Feb. 7, 1861; S. Wyatt & Co. to Sutherlin, Feb. 9, 1861; both, Folder 9, Box 2, William Thomas Sutherlin Papers, SHC; James M. McPherson, *Battle Cry of Freedom* (Oxford University Press, 1988), 274, 278–80; William H. Gaines, Jr., *Biographical Register of Members, Virginia State Convention of 1861* (Virginia State Library, 1969), 39, 74.

2. Norrell, McLaughlin & Co. to W. T. Sutherlin, Dec. 3, 1860, Folder 8, Box 1, Sutherlin Papers.

3. Peter McEnery to W. A. & G. Maxwell, May 9, 1861, letterbook of Peter McEnery, Jan. 14, 1861–Jan. 8, 1864, Peter McEnery Papers, VHS (hereafter McEnery letterbook 1861–1864). See also McEnery to C. R. Somervail & Co., May 3, 1861; and McEnery to D. H. Watkin & Co., May 3, 1861, in the same source. David G. Surdam, "Northern Naval Superiority and the Economics of the American Civil War," *JEH* 56 (June 1996): 473–75.

4. Peter McEnery to James Hunter & Co., June 14, 1861; McEnery to C. R. Somervail & Co., Mar. 12, 1862; both, McEnery letterbook 1861–1864.

5. John M. Sutherlin to W. T. Sutherlin, Aug. 20, 1861, Folder 12, Box 2, Sutherlin Papers.

6. Christian & Lea to W. T. Sutherlin, Jan. 3, 1862, Folder 15, Box 2; Coleman Wortham to Sutherlin, May 1, 1862, Folder 18, Box 2; both, Sutherlin Papers. For slave-hiring practices, see Peter McEnery, "Tobacco Factory Daybook," Dec. 1837–Jan. 1857, pp. 34, 54, 95–98, 129, s.vv. June 2, 1838, Sept. 8, 1838, Jan. 3, 1839, Mar. 25, 1840, Thomas Branch & Co. Papers.

7. "Auction #2," Nov. 4, 1862, Auction Sales Record Book, 1854–1867, Thomas Branch & Co. Papers; Peter McEnery to Brandon Chambliss & Co., Mar. 13, 1862, McEnery letterbook 1861–1864.

8. James Thomas Butler, "Diary," s.vv. Oct. 26, 1862, Jan. 22. 1863, Feb. 7, 1863, VHS.

9. Hill & Warren to Samuel Ayres & Son, Jan. 14, 1862, Folder 15, Box 2; Hill & Warren to Sutherlin, Jan. 20, 1862, Folder 15, Box 2; both, Sutherlin Papers.

10. C. Dimmock to Robert E. Lee, Apr. 27, 1861, in U.S. War Department, *War of the Rebellion . . . Official Records*, series 1, 53 vols. (GPO, 1880–1901), 1:784–85 (hereafter O.R.); J. A. Early to R. S. Garnett, May 16, 1861, O.R. 1:851–53; Nelson Lankford, *Richmond Burning* (Viking, 2002), 19.

11. Richard Franklin Bensel, *Yankee Leviathan* (Cambridge University Press, 1990).

12. Joseph C. G. Kennedy, [*Census of*] *Agriculture of the United States in 1860* (GPO, 1864), 63, 163; Francis A. Walker, [*Census of*] *Statistics of the Wealth and Industry of the United States* (GPO, 1872), 3:83.

13. Tobacco Institute, "Kentucky's Tobacco Heritage," pamphlet, Folder 333, Box 4, Series 4.4, Universal Leaf Tobacco Company Papers, VHS; J. B. Killebrew, "Report on the Culture and Curing of Tobacco in the United States," in *Report on the Productions of Agriculture as Returned at the Tenth Census*, vol. 3 (GPO, 1883), 668–70.

14. Secretary of the Treasury, "Tobacco," May 28, 1842, H.doc. 235 (27–2) 404, p. 5.

15. Killebrew, "Report on the Culture and Curing," 704; W. S. Kline & Company, "Greater Durham in 1904-05," pamphlet (Seeman Press, 1905), 1, PLSC; William Clark to James Clark, Mar. 4, 1863, Folder 3, Correspondence 1860–1864, Clark-Strater-Watson Family Papers, FHS.

16. Lankford, *Richmond Burning*, 25, 36, 60–65, 85–86, 90, 93, 104; *Richmond Dispatch*, Dec. 11, 1889.

17. Lankford, *Richmond Burning*, 2–3, 56, 104, 137.

18. Claudia Goldin and Frank Lewis, "The Economic Cost of the American Civil War: Estimates and Implications," *JEH* 35 (June 1975): 299–326; Charles A. Beard and Mary R. Beard, *The Rise of American Civilization*, 2 vols. (Macmillan, 1927); Ralph Andreano, ed., *The Economic Impact of the Civil War* (Schenkman, 1962); Bensel, *Yankee Leviathan*; Paludan, "What Did the Winners Win?" in James M. McPherson and William J. Cooper, Jr., eds., *Writing the Civil War* (University of South Carolina Press, 1998), 174–200.

19. R. G. Dun & Co., *Virginia*, 43:101, s.v. William Greanor & Son, R. G. Dun & Co. Collection, HBS.

20. James M. McCulloch, "Ledger, 1861–1865," 55, PLSC; "Description and Guide," Thomas Branch & Co. Papers, 2–3.

21. S. Wyatt & Co. to W. T. Sutherlin, Feb. 18, 1861, Folder 9, Box 2, Sutherlin Papers; R. G. Dun & Co., *North Carolina*, 11:505/11, 54, s.v. Robert A. Jenkins, R. G. Dun & Co. Collection; Charles B. Ball to W. T. Sutherlin, Sept. 14, 1865, Folder 23, Box 3; "Bought of P. S. March," receipt, Oct. 31, 1865; Folder 23, Box 3; both, Sutherlin Papers.

22. Solon McElroy to Robert A. Lancaster, Apr. 10, 1865, Lancaster Family Papers, VHS.

23. Bensel, *Yankee Leviathan*, ix, 2; Richard R. John, *Network Nation* (Harvard University Press, 2010).

24. John H. Garber, "Tobacco," in U.S. Bureau of the Census, *Twelfth Census of the United States*, vol. 9, *Manufactures*, pt. 3 (GPO, 1902), 671; W. Elliot Brownlee, *Federal Taxation in America* (Cambridge University Press, 1996), 9, 13–16, 23, 26, 27, 29.

25. Wilbur R. Miller, *Revenuers and Moonshiners* (University of North Carolina Press, 1991), 6, 68, 139–40, quotation at 168; Alfred D. Chandler, Jr., *The Visible Hand* (Harvard University Press, 1977); Naomi Lamoreaux, *The Great Merger Movement in American Business, 1895–1904* (Cambridge University Press, 1985).

26. Annual Report of the Commissioner of Internal Revenue for Fiscal Year ending June 30, 1868, H.exdoc. 5 (40–3) 1371, p. 3; Annual Report of the Commissioner of Internal Revenue for Fiscal Year ending June 30, 1866, H.exdoc. 4 (39–2) 1291, pp. 7, 11; Annual

Report of the Commissioner of Internal Revenue for Fiscal Year ending June 30, 1867, H.exdoc. 5 (40–2) 1329, pp. 3, 8; Garber, "Tobacco," 671.

27. Edward Burke, *Tobacco Manufacture in the United States* (American News Company, [1864]), 3, 5; Garber, "Tobacco," 671–72; "Resolutions of the Legislature of Kentucky, in relation to the proposed tax on tobacco," Mar. 24, 1862, S.misdoc. 68 (37–2) 1124; Ohio Legislature, "Objections," Feb. 13, 1866; S.misdoc. 64 (39–1) 1239; "Concurrent Resolution," Feb. 11, 1865, S.misdoc. 45 (38–2) 1210.

28. Burke, *Tobacco Manufacture*, 5; *Report of the Industrial Commission on Agriculture*, vol. 11 (GPO, 1901), 51–71.

29. Annual Report of the Commissioner of Internal Revenue for . . . 1867, 8; Annual Report of the Commissioner of Internal Revenue for Fiscal Year ending June 30, 1870, H.exdoc. 4 (41–3) 1452, p. 9.

30. Annual Report of the Commissioner of Internal Revenue for . . . 1870, 9.

31. Ibid., 10.

32. Ibid., 10–11; Annual Report of the Commissioner of Internal Revenue for Fiscal Year ending June 30, 1872, H.exdoc. 4 (42–3) 1563, pp. 11, 12; the Annual Reports of the following years indicate the increasing success of this effort: see, for example, H.exdoc. 4 (43–1) 1604; H.exdoc. 4 (43–2) 1642.

33. Annual Report of the Commissioner of Internal Revenue for Fiscal Year ending June 30, 1873, H.exdoc. 4 (43–1) 1604, pp. 9–10; Annual Report of the Commissioner of Internal Revenue for Fiscal Year ending June 30, 1871, H.exdoc. 4 (42–2) 1508, p. 10; Annual Report of the Commissioner of Internal Revenue for Fiscal Year ending June 30, 1877, H.exdoc. 4 (45–2) 1804, p. 15.

34. Annual Report of the Commissioner of Internal Revenue for . . . 1866, 11–12; "Resolution," Mar. 2, 1882, S.misdoc. 61 (47–1) 1993; S.misdoc. 7 (43–2) 1630; R. G. Dun & Co., *North Carolina*, 10:467, s.v. Rufus H. Bobbitt, R. G. Dun & Co. Collection.

35. R. G. Dun & Co., *North Carolina*, 8:407E, 8:416, s.vv. D. L. Dyson; R. G. Dun & Co., *Virginia*, 44:221, s.v. J. B. Pace & Co., R. G. Dun & Co. Collection.

36. George A. Akerlof, "The Market for 'Lemons,'" *QJE* 83 (Aug. 1970): 488–500.

37. R. G. Dun & Co., *Virginia*, 44:283 s.vv. Thomas J. Noble, Ed C. Mayo, and T. J. Noble; Wiebe E. Bijker, Thomas P. Hughes, and Trevor J. Pinch, eds., *The Social Construction of Technological Systems* (MIT Press, 1987), introduction; Susan Leigh Star and James R. Griesemer, "Institutional Ecology, 'Translations,' and Boundary Objects," *Social Studies of Science* 19, no. 3 (Aug. 1989): 387–420; Martha Lampland and Susan Leigh Star, *Standards and Their Stories* (Cornell University Press, 2009).

38. R. G. Dun & Co., *North Carolina*, 11: 504D/20, s.v. Erastus Mitchell, R. G. Dun & Co. Collection.

39. R. G. Dun & Co., *Virginia*, 43:163, 43:370/400, 43:476, s.vv. Ballanff & Co., Jno. F. Allen & Co. (L. Ginter), and John F. Allen & Co., R. G. Dun & Co. Collection; John Pope to B. N. Duke, Mar. 14, 1891, Folder 1891 March (Corr), Box 1, Benjamin Newton Duke Papers, PLSC.

40. Chandler, *Visible Hand*, 1, 81, 249, 290, 382; R. G. Dun & Co., *North Carolina*, 19:228, s.v. Wm. I. Duke & Co. (which refers to Washington Duke and his brief partnership with Francis Stagg), R. G. Dun & Co. Collection; analyzing Washington Duke's political affilia-

tions supports this claim of his business, not subsistence, orientation: see the careful and accurate treatment of his politics in Robert F. Durden, *The Dukes of Durham, 1865–1929* (Duke University Press, 1987), 6–8, 27; Nannie May Tilley, *The Bright-Tobacco Industry* (University of North Carolina Press, 1948), 555–57, 593–98.

41. Lamoreaux, *Great Merger Movement*, 28; Walter Licht, *Industrializing America* (Johns Hopkins University Press, 1995), 127, 145; *Tobacco (London)*, vol. 9, no. 103 (1889), 198, quoted in Howard Cox, *The Global Cigarette* (Oxford University Press, 2000), 47.

42. Leslie Hannah, "Whig Fable of American Tobacco," *JEH* 66 (Mar. 2006), 64; Cox, *Global Cigarette*, 53; Thomas P. Hughes, *Networks of Power* (Johns Hopkins University Press, 1983); Donald Finlay Davis, *Conspicuous Production* (Temple University Press, 1988); Tilley, *Bright-Tobacco Industry*, 576; Lewis K. Keizer to R. L. Patterson, July 9, 1894, Folder 1, Rufus Lenoir Patterson Papers, SHC; J. M. Currin to B. N. Duke, Sept. 9, 1891, Folder 1891 Sept. 1–10 (Corr), Box 1, Benjamin Newton Duke Papers.

43. Department of Commerce and Labor, *Report of the Commissioner of Corporations on the Tobacco Industry*, 3 vols. (GPO, 1909), 1:2, 2:40; Lamoreaux, *Great Merger Movement*, table 1.2, demonstrates that the ATC needed separate firms to control different parts of the market, including snuff, cigars, and stogies. The Continental Tobacco Company—the ATC's plug branch—was formed in 1898.

44. Commerce and Labor, *Report on the Tobacco Industry*, 1:94–95, 1:223.

45. Ibid., 1:74, 1:85, 1:95, 1:223; "Free Tobacco Bill," Mar. 1, 1907, S.doc. 372 (59–2) 5073, pp. 10–11, 131, 156; Secretary of the Treasury, "Tobacco," May 28, 1842, H.doc. 235 (27–2) 404, p. 5.

46. Commerce and Labor, *Report on the Tobacco Industry*, 1:2, 1:94, 1:309; Cox, *Global Cigarette*, 23–24, 59; Jarrett Rudy, *The Freedom to Smoke* (McGill-Queen's University Press, 2005), 3; Allan M. Brandt, *The Cigarette Century* (Basic Books, 2007), 43.

47. Commerce and Labor, *Report on the Tobacco Industry*, 2:xx, 2:27, 1:74, 1:140.

48. "Bill of Complaint and Answer in B. F. Gravely & Sons vs. B. F. Gravely & Co.," Nov. 24, 1888, Folder 662, Box 20, pp. 7, 9; "Tobacco Book," 1860[–1862], Item 29, Box 46; "The Deposition of P. B. Gravely taken by defendants at Danville," Oct. 24, 1889, pp. 1–2; Folder 662, Box 20; both, Gravely Family Papers, LOV.

49. "Bill of Complaint and Answer," 2, 13.

CHAPTER 4: RIPENESS IS ALL

1. Gavin Wright, *Old South, New South* (Louisiana State University Press, 1986); Harold D. Woodman, *New South, New Law* (Louisiana State University Press, 1995); Orville Vernon Burton and Robert C. McMath, eds., *Toward a New South?* (Greenwood Press, 1982); for eighteen-month production cycle, see Frederick F. Siegel, *Roots of Southern Distinctiveness* (University of North Carolina Press, 1987), 62.

2. L. J. Bradford (Augusta, KY), "The Culture and Management of Tobacco," *Report of the Commissioner of Agriculture for the Year 1863* (GPO, 1863), 87–91; Walter W. W. Bowie (Prince George's County, MD), "Culture and Management of Tobacco," *Report of the Commissioner of Agriculture for the Year 1867* (GPO, 1868), 179–85; the annual *Report* for 1874 is likewise instructive. A. Hunter Dupree, *Science in the Federal Government* (Harvard Univer-

sity Press, 1957); Charles E. Rosenberg, *No Other Gods* (Johns Hopkins University Press, 1976; rev. ed., 1997); Lawrence Busch and William B. Lacey, *Science, Agriculture, and the Politics of Research* (Westview Press, 1983); Daniel Carpenter, *The Forging of Bureaucratic Autonomy* (Princeton University Press, 2001); G. Terry Sharrer, *A Kind of Fate* (Iowa State University Press, 2000), 115. A growing body of research on varietal types sees the property rights developing in this period as crucial: Glenn Bugos and Daniel J. Kevles, "Plants as Intellectual Property," *Osiris* 2nd ser. 7 (1992): 74–104; Daniel J. Kevles, "Patents, Protections, and Privileges," *Isis* 98 (June 2007): 323–31.

3. Ira Berlin et al., eds., *Slaves No More* (Cambridge University Press, 1992), 72–73. David W. Blight debunks this scene in *A Slave No More* (Mariner Books, 2009).

4. "The Labor System of the South," *Southern Cultivator and Dixie Farmer* 36 (Nov. 1878): 427.

5. Sprigg Russell, affidavit before the Justices of the Peace of Granville County, n.d., Miscellaneous Records, Oxford, North Carolina, BRFAL, NARA Microfilm Publication M1909, roll 48.

6. Tho. H. Hay? to Col. Bomford?, Jan. 25, 1868, Feb. 25?, 1868, monthly reports, Letters Sent, Henderson, North Carolina, BRFAL, NARA Microfilm Publication M1909, roll 23.

7. William Kauffman Scarborough, *Masters of the Big House* (Louisiana State University Press, 2006), 197–200.

8. "Thoughts for the Month," *Southern Cultivator and Dixie Farmer* 34 (Nov. 1876): 420; see also vol. 35 (Jan. 1877): 3, to see the paternalist-planter cast to this advice.

9. See, for example, "Ledger," 1876, Folder 160.25.a; and "General Records and Accounts," 1885–1909, Folder 160.27.a; both, Elias Carr Papers, ECU.

10. Roger L. Ransom and Richard Sutch, *One Kind of Freedom* (Cambridge University Press, 1977), 117, 121–29, 137–48; Scott Marler, "Merchants and the Political Economy of Nineteenth-Century Louisiana: New Orleans and Its Hinterlands" (PhD diss., Rice University, 2006).

11. Woodman, *New South, New Law*, 87–94.

12. Wm. Jervis, "Complaint [of Howard Thomas]," May 11, 1867, Contracts and Transcripts of Court Cases, Oxford, North Carolina, roll 48; Thos. Hay to N. R. Jones, Nov. 1, 1867, Jan. 8, 1868, Letters Sent, Henderson, North Carolina, roll 23; all, BRFAL, NARA Microfilm Publication M1909.

13. Thos. Hay to Jacob F. Chur?, Nov. 25, 1867; Thos. Hay to N. R. Jones, Jan. 8, 1868; both, Letters Sent, Henderson, North Carolina, BRFAL, NARA Microfilm Publication M1909, roll 23.

14. A. V. Sims to J. D. Setliffe, Apr. 13, 1903, Apr. 23, 1903, May 12, 1903, Folder 243, Box 25, Wilson and Hairston Family Papers, SHC. It seems possible that hilling was an example of Genovese's law, a means of making regular use of labor when labor was a fixed cost, a capital investment into slave property, and it disappeared as its purposes did. Hilling also likely aided drainage, which fertilizer may have made less necessary.

15. Beth Cannady, "We Own Our Land," 248/14426, oral history interview, Folder 8, Leonard Rapport Papers, SHC; "Harvesting and Curing Tobacco," *Rural Carolinian* 1 (Aug. 1870): 664; "Stalk harvesting Bright Leaf tobacco, B. F. Williamson farm, Darlington, about

1897," in Eldred E. Prince, Jr., with Robert R. Simpson, *Long Green* (University of Georgia Press, 2000), picture inserted after p. 138.

16. Thomas Hoyt, "Improvement in Curing Tobacco-Stems," U.S. Patent No. 7,001, granted Jan. 8, 1850; Nannie May Tilley, *The Bright-Tobacco Industry* (University of North Carolina Press, 1948), 64–68.

17. Martin Hill & Co. to Wm. H. Burwell, Oct. 25, 1879, Folder 49, Box 2, Series 1.2, Burwell Family Papers, SHC.

18. Waltz Maynor, Mar. 12, 2002, oral history interview, in the possession of the author.

19. W. W. Garner et al., "History and Status of Tobacco Culture," in USDA *Yearbook* 1922 (GPO, 1923), 434; Julia A. King, "Tobacco, Innovation, and Economic Persistence in Nineteenth-Century Southern Maryland," *Agricultural History* 71 (Spring 1997): 207–36.

20. "Descriptive of the Modern Tobacco Barn," pamphlet (Enterprise Job Print, 1886); J. B. Hunter, "Useful Information concerning Yellow Tobacco, and Other Crops, as Told by Fifty of the Most Successful Farmers of Granville County, N.C.," pamphlet (W. A. Davis, 1880), quotations at 8, 10; William Wallace White, "Diary," vol. 12, s.vv. Sept. 6–7, Sept. 18, Oct. 1, Nov. 8, and Dec. 3, 1875, William Wallace White Papers, SHC.

21. Charles Horace Hamilton, *Gamblers All*, 220–21, Folder 13, Box 14, Charles Horace Hamilton Papers, NCSU; Roger Biles, "Tobacco Towns," *NCHR* 84 (Apr. 2007): 156–90; Elmo L. Jackson, *The Pricing of Cigarette Tobaccos* (University of Florida Press, 1955), 109–11. Auctions often mediate exchanges in conditions of price uncertainty, but limiting sales to open on particular days in particular markets gave buyers more control over leaf prices.

22. Wilbur Wright Yeargin, "The History of the Tobacco Auction System and the Tobacco Auctioneer," (MALS thesis, Duke University, 1989), 1, 27–28; Siegel, *Roots of Southern Distinctiveness*, 30–31.

23. R. G. Dun & Co., *Virginia*, 37:222, s.v. Keen, Poindexter & Co., entry for June 1878–Sept. 1879, R. G. Dun & Co. Collection, HBS.

24. Central Warehouse, "Receipt[s for R. D. Brown]," May 4, 1893, May 12, 1893, Folder 229, Box 23, Wilson and Hairston Family Papers, SHC.

25. Thomas P. Hughes, "Technological Momentum," in Merritt Roe Smith and Leo Marx, eds., *Does Technology Drive History?* (MIT Press, 1994); Wiebe E. Bijker and John Law, eds. *Shaping Technology/Building Society* (MIT Press, 1992); Bruno Latour, *Reassembling the Social* (Oxford University Press, 2005); Paul A. David, "Path Dependence," *Cliometrica* 1, no. 2 (Apr. 2007): 91–114; R. Boyer, "Technical Change and the Theory of *Régulation*," in Giovanni Dosi et al., *Technical Change and Economic Theory* (Pinter, 1988).

26. Edward J. O'Brien to F. W. Hardwick, June 15, 1899; Geo. J. Long to R. W. Bingham, May 3, 1907; both, Folder 93, Robert Worth Bingham Papers, FHS; Tho. S. West (his mark) to Wm. Shanks, Dec. 11, 1877, Folder 32, Box 3, Shanks Family Papers, SHC; William T. Sutherlin, "Account Book, 1872," s.vv. Doctor Brandon, Bailey Claiborne, Henry Dodson, Peter Moseley, Fannie Skipper, Ann Flippin, Cornelia Jameson, and Aunt Betsy, Folder 47, Box 5, William Thomas Sutherlin Papers; see also Finding Guide, William Thomas Sutherlin Papers, for the range of sectors his enterprise covered. Joseph W. Holliday, "Ledger, 1880–

1881," s.v. Dr. Marion R. Skipper, entries for Sept. 16, Nov. 4, Dec. 10, 1880; idem, "Plantation Time Book, 1888–1896," s.vv. Jan. 21, 1888, Apr. 6, 1889, Joseph W. Holliday Papers, USC.

27. William Faulkner, *The Hamlet* (Random House, 1931; 3rd ed., 1964), 3, 5; Ransom and Sutch, *One Kind of Freedom*, 126–48.

28. Finding Aid, Jesse W. Grainger Papers, ECU.

29. Wallace Bros. to Sam Kramer, Sept. 30, 1891, Folder 1891 Sept. 11–30 (Corr), Box 1; L. Ash to B. N. Duke, Sept. 15, 1891, Folder 1891 Sept. 11–30 (Corr), Box 1; J. C. Moore to B. N. Duke, Sept. 9, 1891, Folder 1891 Sept. 1–10 (Corr), Box 1; E. M. Redd to B. N. Duke, April 28, 1892, Folder 1892 April 21–30 (Corr), Box 2; A. J. Boyd to W. W. Fuller, April 28, 1892, Folder 1892 April 21–30 (Corr), Box 2; all, Benjamin Newton Duke Papers, PLSC.

30. Finding Aid, Jesse W. Grainger Papers, ECU; "For Tobacco Growers! Great Inducements Offered in Virginia, near Petersburg," *Progressive Farmer*, Dec. 2, Dec. 9, Dec. 16, and Dec. 23, 1890; Jan. 27, 1891; always on p. 5.

31. Tilley, *Bright-Tobacco Industry*, 349–53.

32. J. J. Atkinson? to Elias Carr, Sept. 4, 1890, Folder 160.3.i; "Barn No. 2," Sept. 16, 1890, Folder 160.27.c; both, Elias Carr Papers, ECU.

33. O. L. Joyner, "Eastern North Carolina as a Tobacco Producing Section," pamphlet (1898), 24.

34. Hamilton, *Gamblers All*, 4.

35. Jimmy M. Skaggs, *The Great Guano Rush* (Macmillan, 1994), 5; Sharrer, *A Kind of Fate*, 43, provides anecdotes of increased use dating from the 1850s, but the war's interruptions make it hard to trace any continuity in the technological change. Stephen Stoll, *Larding the Lean Earth* (Hill and Wang, 2002), traces efforts to establish self-sustaining agriculture in the antebellum Northeast. However, biological, readily available fertilizers (say, manure) had a tendency to reduce the appeal of tobacco; see Lewis Cecil Gray, *History of Agriculture in the Southern United States to 1860* (Carnegie Institution, 1933), 1:217. For the rapid increase in some Southern states' use of commercial fertilizer in the decade after 1880, then, see Ransom and Sutch, *One Kind of Freedom*, 187, 189, table 9.10.

36. "Description and Guide," 3, Thomas Branch & Co. Papers, VHS; Shepherd W. McKinley, "The Origins of 'King' Phosphate in the New South" (PhD diss., University of Delaware, 2003).

37. Department of the Interior, *Report on the Productions of Agriculture as Returned at the Tenth Census*, vol. 3 (GPO, 1883), 103, 596; Sharrer, *A Kind of Fate*, 67.

38. "Tobacco: The Outlook in America for 1875," pamphlet (Southern Fertilizing Company, n.d.); "The Position Tobacco Has Ever Held as the Chief Source of Wealth to Virginia," pamphlet (Southern Fertilizing Company, 1876), VHS; "Major Ragland's Instructions How to Grow and Cure Tobacco, Especially Fine Yellow," pamphlet (Southern Fertilizing Company, 1885); "Wholesale Catalogue of Reliable Tobacco Seeds," pamphlet (R. L. Ragland Seed Co., 1894); "Tobacco: How to Manage It, from the Plant-Bed to the Warehouse," *Progressive Farmer*, Apr. 21, 1891, p. 1; Apr. 25, 1891, p. 4; Jan. 26, 1892, p. 8; Feb. 16, 1892, p. 8.

39. W. J. C. to A. R. Ledoux, Apr. 5, 1879, in North Carolina Agricultural Experiment Station, *Annual Report for 1879* (Observer, State Printer and Binder, 1879), 53; George S. Bradshaw, "Biography of Captain William Henry Snow, 1825–1902," pamphlet (originally

published in the *High Point News*, Mar. 24, 1921); "Descriptive of the Modern Tobacco Barn," "Snow's Modern Barn System of Raising and Curing Tobacco," pamphlet, 3rd ed. (Press of Isaac Friedenwald, 1890), NCC. W. H. Snow to Duke, Sons & Co, Sept. 19, 1891, Folder 1891 Sept. 11–30 (Corr), Box 1; see also Folder 1893 June 16–30 (Corr), Box 3; all in Benjamin Newton Duke Papers, PLSC.

40. R. L. Soyars to A. V. Sims, Sept. 6, 1898, Folder 234, Box 24; for farm location, see A. V. Sims to Harden Hairston, Jan. 8, 1894, Folder 230, Box 23; Hodnett-Vass-Watson Company to A. V. Sims, July 24, 1900, Folder 239, Box 24; all, Wilson and Hairston Family Papers, SHC; Shepherd Supply Co., advertisement, *South Carolina Tobacconist*, Mar. 3, 1896, p. 1; "Flue Book," 1938, Joseph W. Holliday Papers.

41. "The Sumter Market," *South Carolina Tobacconist*, Mar. 3, 1896, [p. 4], uses the term to describe not only the opening of a particular season, but of all leaf sales in that town; Jackson, *Tobacco Prices*, 109–11, demonstrates how staggering the opening days aided buyers in establishing and limiting leaf prices.

42. J. B. Killebrew, "Report on the Culture and Curing of Tobacco in the United States," in *Report on the Productions of Agriculture as Returned at the Tenth Census*, vol. 3 (GPO, 1883), 704.

43. [Jason G. Guthrie], *Measuring America* (GPO, 2002), 125–26, 131–32; Manuel Castells, *The Rise of the Network Society* (Blackwell, 1996); J. B. Killebrew and Herbert Myrick, *Tobacco Leaf* (Orange Judd, 1897; repr. 1934).

44. Killebrew, "Report on the Culture and Curing," 609, 712.

45. Gray, *History of Agriculture*, 1:236.

46. Killebrew, "Report on the Culture and Curing," 797, 798.

47. Ibid., 787, 799.

48. Ibid., 799–801.

49. Ibid., 794; Laura Edwards, "The Politics of Manhood and Womanhood" (PhD diss., University of North Carolina at Chapel Hill, 1991), 26–84, demonstrated that freedpeople sometimes produced excellent tobacco.

50. Killebrew, "Report on the Culture and Curing," 704, 812.

51. Ibid., 704–6. The issue of cost-benefit analysis under conditions of technological change has been examined by Edward W. Constant II, *The Origins of the Turbojet Revolution* (Johns Hopkins University Press, 1980).

52. Killebrew, "Report on the Culture and Curing," 709–12.

53. Ibid., 675–77.

54. Ibid., 674–75, 678–79.

55. Ibid., 676, 792–93.

56. Hunter, "Useful Information concerning Yellow Tobacco," 8.

CHAPTER 5: INVENTING TRADITION

1. "Report Accompanying HR 5720," H.rp. 1635 (47-1) 2070; Lewis Cecil Gray, *History of Agriculture in the Southern United States to 1860*, 2 vols. (Carnegie Institution, 1933), 1:69–73; "Amending Section 3362 of the Revenue Statutes," H.rp. 2712 (57-1) 4407; Santa Fe Natural Tobacco Company, www.nascigs.com (accessed Feb. 24, 2009).

2. John S. Campbell, "The Perique Tobacco Industry of Louisiana," typescript, Nov. 1971, and "The Perique Tobacco Industry of St. James Parish, La.: A World Monopoly," typescript, n.d., both in the John S. Campbell Papers, NCSU.

3. Eric Hobsbawm and Terence Ranger, eds., *The Invention of Tradition* (Cambridge University Press, 1992); Richard A. Peterson, *Creating Country Music* (University of Chicago Press, 1997); E. Melanie DuPuis and Peter Vandergeest, *Creating the Countryside* (Temple University Press, 1996).

4. Thomas P. Hughes, *Rescuing Prometheus* (Pantheon Books, 1998); "Introduction," in Wiebe E. Bijker, Thomas P. Hughes, and Trevor J. Pinch, eds., *The Social Construction of Technological Systems* (MIT Press, 1987).

5. Thomas J. Misa, "Retrieving Sociotechnical Change," in Merritt Roe Smith and Leo Marx, eds., *Does Technology Drive History?* (MIT Press, 1994), 127–37.

6. Department of Commerce and Labor, *Report of the Commissioner of Corporations on the Tobacco Industry*, 3 vols. (GPO, 1909), 1:222; D. C. Patterson to Mess. W. Duke, Sons & Co., Branch, Sept. 16, 1891, Folder 1891 Sept. 11–30 (Corr), Box 1, Benjamin Newton Duke Papers, PLSC.

7. J. B. Cobb to B. N. Duke, Aug. 31, 1891, Folder 1891 Aug. (Corr), Box 1; "Leaf Department," chart, n.d., Folder 1891 May, Box 1; both, Benjamin Newton Duke Papers.

8. J. T. C. Moore to B. N. Duke, Sept. 15, 1891; Wm. A. Marburg to B. N. Duke, Sept. 23, 1891; both, Folder 1891 Sept. 11–30 (Corr), Box 1, Benjamin Newton Duke Papers.

9. J. B. Cobb to B. N. Duke, Nov. 17, 1891, Folder 1891 Nov. (Corr), Box 1, Benjamin Newton Duke Papers; J. B. Duke to W. B. Hawkins, Sept. 4, 1903, letterbook 1894–1904, Box 5, James Buchanan Duke Papers, PLSC.

10. John Pope to B. N. Duke, Oct. 5, 1891, Folder 1891 Oct. 1–11 (Corr), Box 1, Benjamin Newton Duke Papers. On marketing towns as descriptors, see J. Sherman Porter to H. H. Denhardt, Dec. 8, 1925, Folder 34, Henry H. Denhardt Papers, FHS.

11. Nannie May Tilley, *R. J. Reynolds Tobacco Company* (University of North Carolina Press, 1985), 210–13; American Tobacco Company, "Half and Half Burley and Bright," tobacco box, n.d.; R. J. Reynolds Tobacco Co., "Apple Sun Cured," chewing tobacco in original plastic wrapper, n.d.; both, Evelyn L. Gentry Collection, ECU.

12. Howard Cox, *The Global Cigarette* (Oxford University Press, 2000), 20, 76–77; Commerce and Labor, *Report on the Tobacco Industry*, 1:12, 1:182, 1:193; B. N. Duke to W. W. Fuller, Mar. 2, 1893, letterbook Feb. 25–Dec. 30, 1893, Box 64, Benjamin Newton Duke Papers; Barbara Hahn, "Paradox of Precision," *Agricultural History* 82 (Spring 2008): 220–35.

13. John D. Hicks, *The Populist Revolt* (University of Minnesota Press, 1931); Lawrence Goodwyn, *Democratic Promise* (Oxford University Press, 1976); Steven Hahn, *The Roots of Southern Populism* (Oxford University Press, 1983); Robert C. McMath, Jr., et al., "Roundtable on Populism," *Agricultural History* 82 (Winter 2008): 1–35.

14. "Bright Tobacco: An Old Negro the First to Cure It," *Progressive Farmer*, Apr. 14, 1884, p. 4.

15. Charles E. Gage, "American Tobacco Types, Uses, and Markets," USDA Circular No. 249 (Jan. 1933; rev. ed. GPO, 1942), 19, 58–59.

16. B. C. Yang et al., "Assessing the Genetic Diversity of Tobacco Germplasm," *Annals*

of Applied Biology 150 (published online June 5, 2007), quotation at 394; B. T. Galloway, "Applied Botany: Retrospective and Prospective," *Science* n.s. 16 (July 11, 1902): 49-59, esp. 51.

17. Hicks, *Populist Revolt*, 116-17, 144, 170-74, 233.

18. Elmo L. Jackson, *The Pricing of Cigarette Tobaccos* (University of Florida Press, 1955); William H. Nicholls, *Price Policies in the Cigarette Industry* (Vanderbilt University Press, 1951); "Testimony by J. B. Duke," [Feb. 25], 1908, Folder U.S. vs. American Tobacco Company et al., Box 5, James Buchanan Duke Papers.

19. Naomi Lamoreaux, *The Great Merger Movement in American Business, 1895-1904* (Cambridge University Press, 1985), 157.

20. "Tobacco: How to Manage It, From the Plant-Bed to the Warehouse," *Progressive Farmer*, Feb. 10, 1886, p. 1.

21. L. J. Bradford, "The Culture and Management of Tobacco," in the *Report of the Commissioner of Agriculture for the Year 1863* (GPO, 1863), 87-91; Walter W. W. Bowie, "Culture and Management of Tobacco," in the *Report of the Commissioner of Agriculture for the Year 1867* (GPO, 1868), 179-85; on the other hand, the Report of the Statistician in the *Report of the Commissioner of Agriculture for the Year 1874* (GPO, 1875), 42-59, does make reference to "Varieties and Uses" and attempts to describe unique techniques in different regions—a kind of precursor to the Killebrew report of the 1880 census.

22. "Tobacco," *Progressive Farmer*, Feb. 10, 1886, p. 1; Snow appears to be responsible for the reprint of part of H. B. Battle, "Tobacco Curing by the Leaf Cure on Wire and the Stalk Processes," North Carolina Agricultural Experiment Station Bulletin no. 86, May 2, 1892, as "Experiments Showing the Comparative Value of Tobacco," *Progressive Farmer*, Feb. 9, 1892, p. 8; and "The Comparative Value of Curing Tobacco upon the Stalk and the Leaf Cure upon Wire," *Progressive Farmer*, Aug. 22, 1892, p. 8. These looked like articles but appeared on the advertisement page in the months when farmers were preparing their seedbeds and when they would begin to prime if they adopted the technique. Snow also had arrangements for farmers who bought his barn to sell their leaf for top dollar at particular warehouses; see, for example, "Attention Tobacco Growers! Oxford Is Your Market, We Want Snow's Wire-Cured Tobacco!" *Progressive Farmer*, July 28, Aug. 4, Aug. 11, and Oct. 21, 1891, always on p. 6.

23. "IMPORTANT ALLIANCE NOTICE," Jan. 7, 1890, p. 6; "THE OXFORD MEETING: Official Proceedings," Jan. 21, 1890, p. 1; "For Tobacco Growers! Great Inducements Offered in Virginia, near Petersburg," advertisement, Dec. 2, Dec. 9, Dec. 16, and Dec. 23, 1890; Jan. 27, 1891; always on p. 5; all in *Progressive Farmer*. On religion, see Joe Creech, *Righteous Indignation* (University of Illinois Press, 2006).

24. Goodwyn, *Democratic Promise*, 14-16, 201-11, views bimetallism as a "Shadow Movement" working against the true cause of the People's Party.

25. Mark Wahlgren Summers, *Party Games* (University of North Carolina Press, 2004). To maintain a dual focus on forces and agency, see Bruno Latour, "On Recalling ANT," in *Actor Network Theory and After*, ed. John Law and John Hassard (Blackwell, 1999), 17-19.

26. Finding Guide, Leonidas Lafayette Polk Papers, NCSU.

27. "The Founding of North Carolina State University, 1887," Folder University History, General, 1889-1921, Box 1, Reference Collection, Institutional Histories, NCSU.

28. Charles Dabney, state chemist and university president, personally took a hand in soil improvements that spread Bright Tobacco cultivation into the Sand Hills region of North Carolina that now includes the Pinehurst development. Charles W. Dabney, "Development in the Sandhills of North Carolina," typescript, Folder 284, Box 22, Series 4, Charles W. Dabney Papers, SHC; North Carolina Agricultural Experiment Station, *Annual Reports* for 1881, 1882, and 1887.

29. George Basalla, *The Evolution of Technology* (Cambridge University Press, 1988); Barbara Ann Kimmelman, *A Progressive Era Discipline* (PhD diss., University of Pennsylvania, 1987).

30. L. J.? Fagan to Elias Carr, Aug. 27, 1889; W. F. Tomlinson to S. B. Alexander, Aug. 29, 1889; both, Folder 160.2.i, Elias Carr Papers, ECU.

31. Christopher Waldrep, *Night Riders* (Duke University Press, 1993); Tracy Campbell, *The Politics of Despair* (University Press of Kentucky, 1993).

32. Waldrep, *Night Riders*, 5, 37–48.

33. Ibid., 87–89, 95–98, 146–60, quotation at 79; Campbell, *Politics of Despair*, 46, 48–51, 72, 78–81, 89–92; Elliot Jaspin, *Buried in the Bitter Waters* (Basic Books, 2007), 87–107.

34. Waldrep, *Night Riders*, 1; Campbell, *Politics of Despair*, 18–19, table 3.

35. Campbell, *Politics of Despair*, 27; Suzanne Marshall, *Violence in the Black Patch of Kentucky and Tennessee* (University of Missouri Press, 1994); Cox, *Global Cigarette*; "Free Tobacco Bill," Mar. 1, 1907, S.doc. 372 (59-2) 5073, esp. pp. 116–17.

36. Nannie May Tilley, *The R. J. Reynolds Tobacco Company* (University of North Carolina Press, 1985), 210–13; C. D. Campbell, "Growing and Curing Tobacco, with especial reference to the tobacco suitable for the West African trade," typescript, n.d. [1932–1936], Folder 8, Charles D. Campbell Papers, FHS; "Annual Reports of the President," vols. 1–3, Minute Books 1914–1923, 1923–1927, 1928–1947, Campbell Company Papers, FHS.

37. J. B. Killebrew, "Report on the Culture and Curing of Tobacco in the United States," in *Report on the Productions of Agriculture as Returned at the Tenth Census*, vol. 3 (GPO, 1883), 638–40.

38. See the manuscript census returns from the above-named respondents, organized in Oct. 2007 alphabetically by state and county, in "Tobacco Schedules 1910," Entry 309, Record Group 29, NARA DC.

39. Ibid.

40. "Exhibition of Tobacco for Premiums," Tobacco 1858–1860 Papers, KYLM; N. B. Berger to John Morgan, May 15, 1869, quotations from Aug. 28, 1869; both, Atkeson-Morgan Family Papers, FHS.

41. Marshall, *Violence in the Black Patch*; Creech, *Righteous Indignation*.

42. Campbell, *Politics of Despair*, 13, 68–69.

43. "Negroes Shot and Whipped," *Louisville Courier-Journal*, Mar. 11, 1908, quoted in Jaspin, *Buried in the Bitter Waters*, 95–100, quotation at 99; Robert Penn Warren, *Night Rider* (Random House, 1939), 260, 305–6; Campbell, *Politics of Despair*, 77–93; C. Vann Woodward, *Origins of the New South* (Louisiana State University Press, 1951), 327; James Michael Rhyne, "Rehearsal for Redemption" (PhD diss., University of Cincinnati, 2006), 190–93, 249–51.

44. "Free Tobacco Bill," 35, 38; W. W. Garner et al., "History and Status of Tobacco

Culture," in USDA *Yearbook* 1922 (GPO, 1923), 427, fig. 19; Eugene D. Genovese, "The Medical and Insurance Costs of Slaveholding in the Cotton Belt," *JNH* 45 (July 1960): 141–55; Ralph V. Anderson and Robert E. Gallman, "Slaves as Fixed Capital," *JAH* 64 (June 1977): 24–46; Killebrew, "Report on the Culture and Curing," 653.

45. Jaspin, *Buried in the Bitter Waters*, 88–90.

46. "That's All," *Hopkinsville Kentuckian*, May 22, 1896, p. 8, LOC. Thanks to Bruce Baker.

47. Bill Cunningham, *On Bended Knees* (McClanahan, 1983), 53; J. P. Thompson & Company, "First Relief for Oppressed Black Patch Farmers," advertisement, *Hopkinsville Kentuckian*, Apr. 3, 1909, sec. 2, p. 12, LOC.

48. Robert F. Durden, *The Dukes of Durham* (Duke University Press, 1987), 165–68.

49. Nicholls, *Price Policies*, 3, 207–9; Commerce and Labor, *Report on the Tobacco Industry*; Jackson, *The Pricing of Cigarette Tobaccos*.

50. J. B. Duke to W. C. McChord, May 4, 1903, May 12, 1903, Sept. 5, 1903, May 9, 1904, May 16, 1904; J. B. Duke to W. B. Hawkins, June 2, 1903, June 10, 1903, Aug. 17, 1903, Sept. 4, 1903, Sept. 28, 1903; J. B. Duke to Charles F. Button, Sept. 21, 1903; all, letterbook 1894–1904, Box 5, James Buchanan Duke Papers, PLSC; Commerce and Labor, *Report on the Tobacco Industry*, 2:46.

51. "Testimony by J. B. Duke," 3296, 3392–99; Campbell, "Growing and Curing Tobacco," "Annual Reports of the President," Campbell Company Papers.

CHAPTER 6: STABILIZATION

1. "Review of Contributing Developments," 3; Tobacco Production Control (Flue-Cured) Forms, Original Contracts and Programs, Materials Leading to; Tobacco Production Control 1934–35, Flue-Cured Contract and Program, Originals; Entry 9; RG 145, Agricultural Adjustment Administration Production Control, Tobacco (hereafter AAA); NARA MD.

2. Robert H. Wiebe, *Businessmen and Reform* (Harvard University Press, 1962); idem, *The Search for Order, 1877–1920* (Hill and Wang, 1967); Daniel T. Rodgers, *Atlantic Crossings* (Harvard University Press, 1998); James R. Grossman, *Land of Hope* (University of Chicago Press, 1989); Clare Corbould, *Becoming African American* (Harvard University Press, 2009); Donald Holley, *The Second Great Emancipation* (University of Arkansas Press, 2000); Jon C. Teaford, *The Unheralded Triumph* (Johns Hopkins University Press, 1984); Zane L. Miller, *Boss Cox's Cincinnati* (Oxford University Press, 1968).

3. Committee on Agriculture, "License and Inspection of Warehouses," Jan. 21, 1916, report to accompany H.R. 9419, H.rp. 60 (64-1) 6903, p. 1; W. W. Garner et al., "History and Status of Tobacco Culture," in USDA *Yearbook* 1922 (GPO, 1923), 444–48.

4. "Economic Background," 3; Tobacco Production Control 1936–39, Flue-Cured Program & Contract, Originals; Entry 9, RG 145; AAA. For price data, see Garner et al., "History and Status," 442 (fig. 23). Allan M. Brandt, *The Cigarette Century* (Basic Books, 2007), 51–54, 97–98; Charles E. Gage, "American Tobacco Types, Uses, and Markets," USDA Circular No. 249 (Jan. 1933; rev. ed. GPO, 1942), 10–11, table 1; 16, table 4.

5. Anthony J. Badger, *Prosperity Road* (University of North Carolina Press, 1980), 24; Garner et al., "History and Status."

6. J. B. Cobb to B. N. Duke, Oct. 12, 1891; John Pope to B. N. Duke, Oct. 16, 1891; both, Folder 1891 Oct. 12–29 (Corr); Cobb to Duke, June 23, 1891, Folder 1891 June (Corr); all, Box 1, Benjamin Newton Duke Papers, PLSC.

7. Committee on Agriculture, "License and Inspection of Warehouses"; Secretary of the Treasury, "To Collect and Publish Statistics of Leaf Tobacco," Jan. 15, 1916, H.doc. 584 (64–1) 7098, p. 2.

8. "Tentative U.S. Standard Grades for Flue-Cured Tobacco," typescript, 1925, Folder 312.13.f, Tapp-Jenkins Papers, ECU; "Tobacco Classification and Inspection," June 14, 1934, H.rp. 2001 (73–2) 9776, p. 2. Annie G. Crisp to Lucy Cherry Crisp, Nov. 1, 1916, Folder 154.8.i; Oct. 6, 1920, Oct. 25, 1920, Folder 154.8.j; all, Lucy Cherry Crisp Papers, ECU.

9. USDA, Bureau of Agricultural Economics, "Type Classification of American-Grown Tobacco," USDA Miscellaneous Circular No. 55 (GPO, 1925), foreword; H. S. Yohe, "The Farmer and the United States Warehouse Act," USDA Miscellaneous Circular No. 51 (GPO, Jan. 1926).

10. John Black, A Dictionary of Economics, 2nd ed. (Oxford University Press, 2003), 66, s.v. "commodity."

11. On tuning, see R. Boyer, "Technical Change and the Theory of Régulation," in Giovanni Dosi et al., Technical Change and Economic Theory (Pinter, 1988).

12. "Tobacco Market," Time, Sept. 17, 1934, www.time.com/time/magazine/printout/0,8816,747979,00.html.

13. Anthony J. Badger, The New Deal (Ivan R. Dee, 2002).

14. J. H. Canady to J. B. Hutson, July 26, 1933, Folder Apr. 1, 1933–July 31, 1933; Subject Correspondence Files, 1933–35; Entry 2, RG 145; AAA.

15. C. C. Davis to R. A. Green, July 19, 1933, Folder Apr. 1, 1933–July 31, 1933; Subject Correspondence Files, 1933–35; Entry 2, RG 145; AAA. On commercial involvement, see J. B. Hutson to Henry Wallace, Aug. 30, 1933; Tobacco Production Control (Flue-Cured) Forms, Original Contracts and Programs, Materials Leading to; Tobacco Production Control 1934–35, Flue-Cured Contract and Program, Originals; Entry 9; RG 145, AAA; also J. C. Lanier to Tobacco Board of Trade, South Boston, Virginia, July 25, 1933, Subject Correspondence Files, 1933–35; Entry 2, RG 145; AAA. Badger, Prosperity Road, 38–42.

16. C. C. Davis to R. A. Green, July 19, 1933.

17. J. C. Lanier, "Memorandum to Mr. Frank," Aug. 7, 1933; J. C. Lanier to Tobacco Board of Trade, South Boston, Virginia [circular letter to twenty-eight additional recipients], July 25, 1933; both, Folder Aug. 1, 1933–Aug. 31, 1933; Subject Correspondence Files, 1933–35; Entry 2, RG 145; AAA. Badger, Prosperity Road, 113, 201–7.

18. J. C. Lanier to W. T. Clark [circular letter to nine additional recipients], July 18, 1933; J. M. O'Dowd to James F. Byrnes, July 2, 1933; A. B. Carrington to J. B. Hutson, July 11, 1933; all, Folder Apr. 1, 1933–July 31, 1933; Subject Correspondence Files, 1933–35; Entry 2, RG 145; AAA.

19. Badger, Prosperity Road, 47–57.

20. Pete Daniel, Breaking the Land (University of Illinois Press), 168–69; Badger, Prosperity Road, 73–74.

21. Beth Cannady, "We Own Our Land," 248/14426, oral history interview, Folder 8,

Leonard Rapport Papers, SHC; Dale Newman, "Work and Community Life in a Southern Town," *Labor History* 19 (Spring 1978): 206; on hilling, see chap. 4.

22. C. C. Davis to R. A. Green, July 19, 1933.

23. Coleman Wortham to William Thomas Sutherlin, Mar. 3, 1862, Folder 16, Box 2, William Thomas Sutherlin Papers, SHC; A. Petricolas & Co. to Messrs. Peters & Whitehead, Sept. 16, 1830, Folder 1, Box 1, Floyd L. Whitehead Papers, SHC.

24. W. W. Garner, H. A. Allard, and E. E. Clayton, "Superior Germ Plasm in Tobacco," in USDA *Yearbook* 1936 (GPO, 1937), especially 818–19.

25. B. C. Yang et al., "Assessing the Genetic Diversity of Tobacco Germplasm," *Annals of Applied Biology* 150 (published online June 5, 2007): 393; Jack Ralph Kloppenburg, *First the Seed* (Cambridge University Press, 1988).

26. Waltz Maynor, Mar. 12, 2002, oral history interview, in the possession of the author. Judith A. McGaw has closely analyzed the way gendered divisions of labor in papermaking dictated the course of industrialization, rather than flowing from it, in *Most Wonderful Machine* (Princeton University Press, 1987). Bulk production of paper occurred regularly to take advantage of men first under full-time artisanal contracts and later under the capital recovery strategies machinery required. Finishing the bulk stock to a high standard, however, was done seasonally, as demand required; this remained women's work and unmechanized. Agriculture, however, rewarded family labor. Replacing barn-side women's work with field stringing would simply have shifted labor from one sex to another.

27. Daniel, *Breaking the Land*, 29.

28. U.S. Patents: 2,704,158, granted Mar. 15, 1955, p. 11; 2,702,134, granted Feb. 15, 1955; 2,704,158, granted Mar. 15, 1955; 2,797,827, granted July 2, 1957; 2,797,827, granted July 2, 1957, p. 4; 2,704,158, granted Mar. 15, 1955.

29. U.S. Patents: 2,696,069, granted Dec. 7, 1954; 2,816,411, granted Dec. 17, 1957; 2,834,173, granted May 13, 1958; 2,834,174, granted May 13, 1958; 2,876,610, granted Mar. 10, 1959; 2,876,610, granted Mar. 10, 1959, pp. 15, 17; 2,816,411, granted Dec. 17, 1957, p. 4; 2,834,173, granted May 13, 1958, p. 3; 2,834,174, granted May 13, 1958; 2,696,069, granted Dec. 7, 1954.

30. Austin "Chick" Smith and William Kinsey, oral history interview, Nov. 30, 1994, SOHP interview K–76, restricted, used with the permission of David Ceceslski, SHC; actually, public policy shifted from acreage to poundage allotments from time to time in the decades after the New Deal.

31. U.S. Patents: 2,687,596, granted Aug. 31, 1954; 3,482,379, granted Dec. 9, 1969.

32. Charles K. Mann, "The Tobacco Franchise for Whom?" in William R. Finger, ed., *The Tobacco Industry in Transition* (Lexington Books, D. C. Heath, 1981), 40; Badger, *Prosperity Road*, 204; Barbara Hahn, "Into the Belly of the Beast," *Southern Cultures* 9 (Fall 2003): 25–50.

33. U.S. Patents: 3,095,230, granted June 25, 1963; 3,143,370, granted Aug. 4, 1964; 3,147,033, granted Sept. 1, 1964; John Beck, "Capital Investment Replaces Labor," *Flue-Cured Tobacco Farmer*, Mar. 1968, pp. 6–7.

34. W. H. Johnson to Allegheny Building Units, Aug. 24, 1960, Folder 6, Box 5, College of Agriculture and Life Sciences, Department of Biological and Agricultural Engineering

Records, NCSU; Bob Davis and Loren A. Ihnen, "An Analysis of Labor Use for Alternative Flue-Cured Tobacco Harvesting and Curing Systems," N.C. Agricultural Experiment Station and USDA, Economics Research Report No. 16 (Sept. 1971). In 1972, when only 1 percent of bright acreage came under the cutters of a mechanical harvester, already 8 percent would undergo bulk curing. In 1979, when between 19 and 33 percent was harvested by machine, already 61 percent was bulk-cured, and the USDA predicted 100 percent bulk-cured product by 1985. Charles Pugh, "Landmarks in the Tobacco Program," 35; Robert Dalton, "Changes in the Structure of the Flue-Cured Tobacco Farm," 64; both in Finger, *Tobacco Industry in Transition*, xi; Daniel, *Breaking the Land*, 263–64.

35. Pugh, "Landmarks in the Tobacco Program," 34, and idem, "The Federal Tobacco Program," 16, 24; both in Finger, *Tobacco Industry in Transition*.

36. Jack O'Keefe, "Thinks Highly of Automatic Primer and Bulk Curing," *Flue Cured Tobacco Farmer*, Jan. 1969, pp. 8–9; John Beck, "Capital Investment Replaces Labor," *Flue Cured Tobacco Farmer*, Mar. 1968, pp. 6–7, in which the picture caption reads, "Helping Ferrell unload a rack of cured tobacco are his son, Hank, and his daughter, Elaine."

37. "A Systems Approach to the Mechanical Harvesting of Bright-Leaf Tobacco," schematics and other papers, Folder 12, Box 5, College of Agriculture and Life Sciences, Department of Biological and Agricultural Engineering Records, NCSU; Robert W. Wilson, "Tobacco Harvester," U.S. Patent 3,083,517, granted Apr. 2, 1963; Robert W. Wilson, Oct. 24, 2002, oral history interview, in the possession of the author; Jesse R. Pinkham, "Method and Apparatus for Harvesting Tobacco," U.S. Patent 3,603,064, granted Sept. 7, 1971; Jesse R. Pinkham, Arthur G. Cockman, and Jerry Ray Joyce, U.S. Patent 3,841,071, granted Oct. 15, 1974.

38. Daniel, *Breaking the Land*, 265.

39. J. C. Trulove, "Tobacco Harvester," U.S. Patent 1,629,422, granted May 17, 1927, p. 3; "Introduction," xi; and Dalton, "Changes in the Structure of the Flue-Cured Tobacco Farm," 63; both in Finger, *Tobacco Industry in Transition*.

40. For an introduction to the debates on economic causes of technological change in agriculture, see Alan L. Olmstead and Paul W. Rhode, "Induced Innovation in American Agriculture," *Journal of Political Economy* 101 (Feb. 1993): 100–118.

APPENDIX. THE REAL THING: TOBACCO GENETICS

1. Loren H. Rieseberg, Troy E. Wood, and Eric J. Baack, "The Nature of Plant Species," *Nature* 440 (Mar. 2006): 524–25.

2. Jan Golinski, *Making Natural Knowledge* (Cambridge University Press, 1998); Ian Hacking, *The Social Construction of What?* (Harvard University Press, 1999).

3. B. C. Yang et al., "Assessing the Genetic Diversity of Tobacco Germplasm," *Annals of Applied Biology* 150 (published online June 5, 2007): abstract, 393, 398, 399.

4. Denis J. Murphy, *People, Plants, and Genes* (Oxford University Press, 2007), 60, 60nn219–21, 62–63; H. Allen Orr and Sarah P. Otto, "Does Diploidy Increase the Rate of Adaptation?" *Genetics* 136 (Apr. 1994): 1475–80.

5. A. D. Shamel and W. W. Cobey, "Tobacco Breeding," USDA Bureau of Plant Industry Bulletin No. 96 (GPO, 1907).

6. W. K. Collins and S. N. Hawks, Jr., *Principles of Flue-Cured Tobacco Production*, textbook (NC State University, 1993), 52–53. I am indebted to Bill Collins, the author of this NC State University textbook, for providing me with his book, which has proved most helpful over the years. See also Sandra Knapp, Mark W. Chase, and James J. Clarkson, "Nomenclatural Changes and a New Sectional Classification in *Nicotiana* (Solanaceae)," *Taxon* 53 (Feb. 2004): 73–82, although the scientists' suggested changes refer only to species other than the one considered in this text, the *N. tabacum* cultivated for commercial purposes.

7. Barbara Ann Kimmelman, *A Progressive Era Discipline* (PhD diss., University of Pennsylvania, 1987); Staffan Müller-Wille, "Leaving Inheritance Behind"; Christophe Bonneuil, "Producing Identity, Industrializing Purity"; both in Max Planck Institut für Wissenschaftsgeschichte, *A Cultural History of Heredity IV*, Preprint 343 (2008); Robert E. Kohler, *Lords of the Fly* (University of Chicago Press, 1994); Edmund Russell, "Evolutionary History," *Environmental History* 8 (Apr. 2003): 204–28.

8. Kohler, *Lords of the Fly*, 29, 32, 45, 79; H. K. Hayes and E. G. Beinhart, "Mutation in Tobacco," *Science* new ser. 39, no. 992 (Jan. 2, 1914): 35.

9. W. W. Garner, "Some Observations on Tobacco Breeding," *Proceedings of the Meeting of the American Breeders Association*, vols. 7–8 (1912), 460–62; Shamel and Cobey, "Tobacco Breeding," 10, 15, 17.

10. T. H. Goodspeed and R. E. Clausen, "Variation of Flower Size in Nicotiana," *Proceedings of the National Academy of Sciences of the USA* 1 (June 15, 1915): 334–35, 337. Experiments with tobacco also provided the general knowledge that plant reproduction is timed by the length of night. Karl C. Hamner and James Bonner, "Photoperiodism in Relation to Hormones as Factors in Floral Initiation and Development," *Botanical Gazette* 100 (Dec. 1938): 388–431.

11. W. A. Setchell, T. H. Goodspeed, and R. E. Clausen, "A Preliminary Note on the Results of Crossing Certain Varieties of Nicotiana Tabacum," *Proceedings of the National Academy of Sciences of the USA* 7 (Feb. 15, 1921): 51, 55.

12. R. Douglas Hurt, *Indian Agriculture in America* (University Press of Kansas, 1987), 58.

13. Robert V. Bruce, *The Launching of Modern American Science, 1846–1876* (Alfred A. Knopf, 1987), 66; Alfred W. Crosby, Jr., *The Columbian Exchange* (Greenwood, 1972); Tod F. Stuessy, *Plant Taxonomy* (Columbia University Press, 1990), 44.

14. Carolus Linnaeus, *Critica Botanica* (Leiden, 1737; trans. Arthur Hort, The Ray Society, 1938), numbers 260 and 262, pp. 123–24, 130–48; "Appendix IV: Selective List of Linnaeus's Works Up to 1753," in *Linnaeus' Philosophia Botanica*, trans. Stephen Freer (Oxford University Press, 2003), 360.

15. Linnaeus, *Critica Botanica*, 41, 55–56, 60; John H. Garber, "Tobacco," in U.S. Bureau of the Census, *Twelfth Census of the United States*, vol. 9, *Manufactures*, pt. 3 (GPO, 1902), 669.

16. Stuessy, *Plant Taxonomy*, 207, 156–57; Jennings, *Practical Treatise*, 34–35.

17. Jackson, *Botanic Terms*; Pamela S. Soltis, Distinguished Professor of Biology at the University of Florida, past president of the Botanical Society of America, and curator of the Laboratory of Molecular Systematics and Evolutionary Genetics at the Florida Museum of Natural History, flirted with the "race" term in a talk delivered at Texas Tech University in 2007. In conversation, a botany professor at my university once said to me on the subject of race, "What are we, pandering? We are *scientists*."

Essay on Sources

The source base for tobacco is enormous. Any keyword search among primary or secondary sources will yield publications and collections numerous and voluminous enough to support an immoderate claim that the story of tobacco is the story of America. The quantity of records required that I choose an approach, a plan of attack. My research began in spring 2002 with certain limits in place: only Bright Tobacco between the Civil War and the New Deal. If a document called from the stacks seemed to discuss burley, for example, it went back. Yet, it was surprising how unspecific most sources were. With not many real meaty references to "Bright Tobacco" appearing in the historical record, the need to perform research on the project I had proposed drove the search in several directions. In the Southern Historical Collection in Chapel Hill, the postbellum Burwell family papers illustrated that varietal types did not yet exist, but rather were under construction. In spring 2003, the R. G. Dun records at the Harvard Business School unveiled the antebellum tobacco industry. That summer, a forced chronological march through the congressional Serial Set provided ever more support for the absence of types in the nineteenth century and the new borders between agriculture and manufacturing that were solidifying after the Civil War.

Suddenly it seemed necessary to grasp and explain the antebellum tobacco trade, in which types did not yet exist. Otherwise, how to capture their construction? It seemed clear from the historical record that today's tobacco types came into existence at some point after the war, because it was so obvious that they *did not yet exist* in the antebellum or colonial period. My database swelled with information on how that happened: New Deal records from the National Archives; vast quantities of postbellum pamphlets that taught cultivation and curing techniques, that sold fertilizer and seeds and promoted specific agricultural techniques along the way. Warehouse receipts and census data demonstrated crop spread, while the *Progressive Farmer*, the Farmer's Alliance, and the history of North Carolina State all seemed part of organizing farmers into producers of specific crops that had a political role to play. It took so long to figure out the first half of the book—the sources of colonial cultivation practices, the origins of the tobacco industry, and the effect of the Civil War—that materials from the postbellum decades awaited the revisions that took place after the doctorate was done.

The causal power of large structural shifts loomed over the project from the start.

The changes wrought by emancipation and the novelty of institutions such as tobacco warehouses, agricultural experiment stations, and guano notes (fertilizer loans) slowly helped to sort out the actions of what historians of technology call "relevant social groups" or sometimes "stakeholders"—people with specific interests in the outcomes of particular technological choices. The border that emerged between agriculture and manufacturing as a result of taxation began to seem central to the emergence of specific tobacco types. As curing and marketing methods became something only farmers did, those techniques began to define the types. Yet regulation proved necessary to unite these designations into a rigid whole: a taxonomy of tobacco types that linked together the crop's region of origin, market purposes, and technology of production.

Antebellum and postbellum records comprised different kinds of sources. In the end, this simple fact dictated the interpretation advanced in the book—that changing institutions of trade shaped technological choices, which then themselves began to shape the trade. Many collections spanned the long nineteenth century and therefore captured those changes. Certain planter families, as well as many commercial and manufacturing firms, left documents that captured their lives and their work both before and after the War of Independence, or the War Between the States. In other cases, different sources in the different periods demonstrated that industrial structures had changed. Merchant accounts and correspondence with planters represented the lions' share of antebellum records. These showed how tobacco growers responded to market signals in growing and selling their crops. Cultivation manuals, descriptions in farm diaries, and several generations of distinguished scholarship fleshed out the analysis of agricultural technology from the colonial period through the Civil War. The dwindling volume of merchant records in the postbellum decades, however, demonstrated the changes in the marketing system. Living in North Carolina, which held the remnants of the twentieth-century system of tobacco production, provided clues on where to look for the work formerly performed by merchants.

The manuscript began to seem like the history of structure, in which structure itself—the very basis of actions and perceptions, what Marx called economic base (as opposed to cultural superstructure)—had been constructed over time and in particular historical circumstances. So many elements appeared crucial to causation that it was tricky to organize them all into a book that made sense. For example, the adoption of fertilizer and the rise of warehouse marketing both contributed to the characteristics of Bright Tobacco, but they themselves had roots in the changing credit arrangements of emancipation. Viewing the postbellum system in terms of institutions and actors, in classic history-of-technology form, eventually provided the spine of the book. As postbellum portions of the manuscript fell into first internalist, then externalist chapters, finally the social construction of tobacco types emerged as a story that someone other than the author might be able to follow.

The Baker Library at the Harvard Business School held the R. G. Dun ledgers that shaped the book so deeply at the start. The National Archives in College Park stuffed me full of New Deal records, while on the mall, in the D.C. facility, years later, the farmers' replies to the census bureau in 1909 illustrated what they thought they were growing. Likewise, the South Caroliniana Collection on the campus of the University of South Carolina provided vivid evidence of the crop spreading while emerging as a distinct category in the trade. So too did the North Carolina State Archives, where a storekeeper's account books demonstrated how new was the crop in East Carolina during the 1890s. The Filson Historical Society welcomed the project quite late in its development. There, merchants in the African trade, and their treatment of to-bacco types in their turn-of-the-century account books, revealed the importance of varietals that grew up together—how the differences between bright and burley and Perique shaped one another, mutually constituting one another in their differences and similarities. The sheer absence of records on the Black Patch was itself reveal-ing. The Christian County Historical Society beckoned, but after eight years, there was really no time for further research. Much work remains. I hope future research-ers who study agricultural technology will construct such stories for other crops, in many locations.

SECONDARY SOURCES

At the start, this project fell mostly within two historical subdisciplines: business his-tory and the history of technology. The postbellum tobacco story—in which a mecha-nizing, consolidating industry drew its raw materials from small farms and increas-ingly artisanal hand-production methods—seemed to contradict dominant theories in each of these fields. Business history asked the question posed in the introduction: why did the industry mechanize and agriculture not? Alfred Chandler's view of verti-cal integration in *The Visible Hand* (Belknap Press, 1977) indicated that firms in this period integrated raw material production into their processes. While the tobacco industry served as one of his main examples of Big Business at the turn of the twenti-eth century, he did not examine raw material production (tobacco agriculture) in his analysis. While this remains a lacuna characteristic of the field, business historians since Chandler have revised, elaborated, and contradicted his claims—a good over-view of the literature can be found in Richard R. John, "Elaborations, Revisions, Dis-sents," in *BHR* 71 (Summer 1997). Alternative interpretations have also added to the picture of business in this period: Philip Scranton, *Proprietary Capitalism* (Cambridge University Press, 1983), added very different business practices that persisted along-side those Chandler examined. For a strong synthesis of more recent developments, see Naomi R. Lamoreaux, Daniel M. G. Raff, and Peter Temin, "Beyond Markets and Hierarchies," *AHR* 108 (Apr. 2003).

Likewise, in the history of technology, the place of raw material production seemed beyond the interest of dominant theories. Most historians of technology fight technological determinism, and a major advance into its lines was best articulated by Thomas Hughes in *Networks of Power* (Johns Hopkins University Press, 1983). Hughes used the case study of electrification to examine the development of large-scale tech-nological systems. In Hughes's systems theory, bottlenecks (known as "reverse sa-lients") drew the attention of engineers until they were fixed. Thus, Thomas Edison's invention of the lightbulb became, rather than an individual act of invention that sparked electrification, a part of the construction of a much larger system, ranging from hydroelectric dams and distribution circuits to lightbulbs in private homes. As in business history, systems theory indicated that small-scale artisanal agriculture should get worked into the industrial processes of high-volume production. That did not happen. Therefore, the border between tobacco agriculture and tobacco manufacturing raised a question of systems theory: tobacco's disparate agricultural and industrial sectors asked why raw material production so rarely seemed a bottleneck. Tungsten for lightbulbs is still mined in primitive conditions, after all. Gabrielle Hecht's examination of French nuclear power (*The Radiance of France*, MIT Press, 1998), which touched on the uranium supplied by France's African colonies, provided clues. What stands inside a large-scale technological system, and what lies outside it? How did those borders first appear?

This question led me to commodity history, which only solidified my sense of how much the history of technology, and its resistance to technological determinism—our interest in the causes of technological change, as well as its effects—could contribute to traditional fields. Older scholarship on commodities provided rich information but lacked the agency and contingency demonstrated in case after case by historians of technology. In both economic-history and popular-history treatments of commodities, crop determinism marred historical analysis. Supply and demand too often looked like products of the crop's own inherent characteristics, or of traits achieved by lucky chance or discovery. Texts included E. R. Billings, *Tobacco* (American, 1875); Charles L. Bartlett, *Guano* (C. C. P. Moody, 1860); William Harrison Ukers, *All About Coffee* (Tea and Coffee Trade Journal Company, 1922); Joseph Clarke Robert, *The Tobacco Kingdom* (Duke University Press, 1938); Nannie May Tilley, *The Bright-Tobacco Industry, 1860–1929* (University of North Carolina Press, 1948); and Arthur Pierce Middleton, *Tobacco Coast* (Mariners' Museum, 1953). Commodity studies, almost infinite in number, formed the backbone of an earlier generation's economic history. Their contributions provide excellent analysis. However, their approaches have become outdated from the perspective of historians of technology who generally see much more chance and human effort at work in the discovery and use of even natural goods.

More recent commodity chain studies, on the other hand, contributed a great deal to my perception of the sources. The scholarship emerging from sociology and globalization studies especially helped explain the interactions between agricultural techniques and the markets for their products: Terence K. Hopkins and Immanuel Wallerstein introduced the field in "Commodity Chains in the World-Economy Prior to 1800," *Sociological Review* 10 (Summer 1986). Collections of essays fleshed out the domain: Gary Gereffi and Miguel Korzeniewicz, eds., *Commodity Chains and Global Capitalism* (Greenwood Press, 1994); Alex Hughes and Suzanne Reimer, *Geographies of Commodity Chains* (Routledge, 2004). In addition, borrowing network theories from sociology helped me extend these findings to human actions: Mark Granovetter, "The Strength of Weak Ties," *American Journal of Sociology* 78 (May 1973), and Manuel Castells, *The Rise of the Network Society* (Blackwell, 1996), provided rich food for thought in the social construction of seemingly natural phenomena. Giovanni Arrighi's work on hybrid forms of capitalism, too (*The Geometry of Imperialism*, Milan: Feltrinelli, 1978; Verso, 1983), joined some of the theories proposed by development economists such as Debraj Ray in *Development Economics* (Princeton University Press, 1998) to apprehend the importance of the agricultural-industrial relationship. Historians of technology have long borrowed from sociologists, and these scholars helped me grasp the historical emergence of the world that now seems natural.

In recent years, too, popular treatments of commodities have boomed. In almost all these cases, however, the story has a particular shape—one familiar to historians of technology and long abandoned by experts in our field. Too often, the commodity under consideration is imbued with remarkable powers and serious impact on the world around it. Mark Kurlansky followed up *Cod: A Biography of the Fish that Changed the World* (Penguin, 1997) with *Salt: A World History* (Penguin, 2002). Iris and Alan MacFarlane explored *The Empire of Tea* (Overlook, 2004), and Larry Zuckerman found *The Potato: How the Humble Spud Rescued the World* (North Point, 1998), while Dan Koeppel studied the *Banana: The Fate of the Fruit That Changed the World* (Plume, 2007). Economic historians have likewise tended to view a commodity as something discovered that then had an effect. The current crop of popular commodity studies, for a general audience, makes more grandiose claims but follows the same line. The very familiarity of this deterministic type of story rings alarm bells for historians of technology. We have spent so long examining not only effects but also causes that, in the case of a commodity, traded between people for specific purposes, the history of technology begs a question: How did people in the past know what it was, the article that lay between them? How did they determine its value?

In *Making Tobacco Bright*, the narrative follows a different form. Instead of identifying a commodity that changed the world, in this case, the world created the commodity. Market relations and regulations shaped tobacco into types, over several centuries. Studying the historical development of those categories into which data themselves are placed makes one critical of analysis that relies on distinctions between agriculture and manufacturing, among plant cultivars, and among all those human-shaped categories that developed in historical contexts. Moreover, for all the importance of technological change in generating economic growth, the approaches developed by generations of historians of technology seem unknown to the popular approach to commodities as well as to economic historians. The diffusion and adoption of technology is a principal subject of interest, but, to a historian of technology, that is very late in the process of technological change to look for its causes. The best way of accomplishing a task appears best only in retrospect. At the time, choices and costs were almost always underdetermined. This remains a significant problem in the popular reliance on market explanations, as well as in economic history scholarship. When teaching business and economic history courses, for example, it is often necessary to correct those places where a textbook simply gets it wrong, owing to ignorance of the history-of-technology field and its findings. Eli Whitney did not invent the cotton gin—not exactly. More important, because longer-established, is the fact that Whitney *did not ever* achieve interchangeable-parts manufacturing. Not even close.

Such mistakes mean little on their own, perhaps, but indicate a field whose members read too little outside its own narrow bounds. To find this error promulgated in current economic history literature, it is probably unfair to cite old texts such as Robert Heilbroner and Aaron Singer, *The Economic Transformation of America, 1600 to the Present*, 4th ed. (Wadsworth Cengage Learning, 1999), and Harold C. Livesay, *American Made* (Longman, 1979). Yet, although these works were first written quite some time ago, they have survived. Still in print and still regularly assigned, they shape interpretations, especially in introductory studies. The alternative case has been available since Merritt Roe Smith published *Harpers Ferry Armory and the New Technology* (Cornell University Press, 1977); see his direct strike against the Whitney legend in Carroll W. Pursell, ed., *Technology in America* (1981; 2nd ed., MIT Press, 1990). On the cotton gin, see Angela Lakwete, *Inventing the Cotton Gin* (Johns Hopkins University Press, 2005). Somewhat better are Jeremy Atack and Peter Passell, *A New Economic View of American History*, 2nd ed. (W. W. Norton, 1994)—yet one should not have to reach that level of training in economic history to learn things that historians of technology have known for more than thirty years. While my book has been heavily influenced by the economic historians with whom I studied, for a long time it was mostly critical of the subdiscipline.

To be entirely frank, there is nothing wrong with quantification. Any tool a historian chooses may get the job done, and statistics are a splendid and revealing device. I have learned so much from agricultural census data and from inserting it into the historical county base maps downloaded from the National Historical Geographic Information System; these provided the maps that appear in this book. However, those who arm themselves with quantitative evidence often claim that some superior rigor derives from their data sets. Yet, the numbers that illuminate historical events have always been first generated and then later compiled by human efforts. Such work lacks perfection and should avoid the pretension of positivism—especially when practitioners choose sides rather than evaluating analysis.

Rather than pick fights and name names, however (as done above in the case of Eli Whitney), I wish instead to acknowledge my debts to the field and point out where in general its practitioners could be more influential. Answering the questions posed by economic historians influenced both the sources here employed and the arguments advanced in these pages. I have been too much concerned with structure to ignore factor endowments—which of course shape history. I would point out, however, that these are not entirely natural. The relative values and costs of land, labor, and capital do themselves have causes—as when labor costs changed as a result of emancipation. Institutions influence costs, and vice versa. As this book paints it, base and superstructure intertwine, and technology occupies a weird middle position between causes and effects. Examples abound in the chapters, to demonstrate that structures have human histories.

Scholars in science and technology studies (STS) have always been borrowers from one another's fields: sociology, anthropology, and history are the major disciplines represented, but economists of innovation and philosophers have always played a role as well. As a cognate discipline, our more humanistic side has long been principally concerned with dismantling or at least questioning technological determinism: the essays collected by Merritt Roe Smith and Leo Marx in *Does Technology Drive History?* (MIT Press, 1994) lay out multiple positions and approaches to the problem. In addition, our ever-expanding subject matter has begun to bring the history of technology into conversation with many other traditional fields. For example, STS scholars have begun to address agricultural technology in greater numbers. Margaret W. Rossiter's *The Emergence of Agricultural Science* (Yale University Press, 1975) and Deborah Kay Fitzgerald's *The Business of Breeding* (Cornell University Press, 1990) have influenced all who work in the field. More recently, Joe Anderson's *Industrializing the Corn Belt* (Northern Illinois University Press, 2009) complements Fitzgerald's work on breeding hybrid corn. Mark Finlay's study of *Growing American Rubber* (Rutgers University Press, 2009) likewise contributes STS approaches to agricultural history. Together these authors have created a narrative of technological change based partly on farmers' choices and partly on the institutions and industries of agricultural science, in both the nineteenth and the twentieth centuries.

Within the field of the history of technology, my approach has followed the evolution of the field for the most part. Systems theory provided some of the basis for the "social construction" model, as established in Wiebe E. Bijker, Thomas P. Hughes, and Trevor J. Pinch, *The Social Construction of Technological Systems* (MIT Press, 1987). Most STS practitioners and historians of technology have accepted at least some of the findings and approaches generated by that school of thought, especially its requirement that historians not assume the outcome, and that we assess successful and failed technologies in symmetrical terms. In addition, the relevant social groups of the social construction generation have become the stakeholders described in Thomas P. Hughes, *Rescuing Prometheus* (Pantheon Books, 1998). By figuring out the interests of participants and their impact on the final result—the success or failure of a technology, its adoption or disappearance or abeyance—STS offers the best means for explaining the contingency behind our devices and behind our systems for dealing with and employing the fruits of the natural world. An excellent recent text that takes these understandings for granted, as the basis of a new exploration, is David Edgerton's *The Shock of the Old* (Oxford University Press, 2006), which moves the global history of technology away from the belief in progress that so animates technological determinism.

Meanwhile, social constructivist scholarship has advanced into actor-network theory, developed in the sociological and anthropological ends of STS. When combined with the systems theory with which this study began, actor-network theory provided the framework for analysis of the primary sources in *Making Tobacco Bright*. Crucial works in the field include John Law and John Hassard, eds., *Actor Network Theory and After* (Blackwell, 1999), and Bruno Latour, *Reassembling the Social* (Oxford University Press, 2005). As Latour has summarized the approach, both social and economic structures have been built by historical actors. To assume the static existence of economic, technological, linguistic, cultural, or social order would leave in place the Marxian assumption that economic base dictates cultural superstructure. Marx also famously declared that "the hand-mill gives you feudalism, the steam-mill gives you industrial capitalism"; this is the very technological determinism that historians of technology most often deny, since little evidence can be marshaled for it in specific case studies. Marx himself knew, after all, that "men make history, but not in conditions of their own making." This book has attempted to contribute to these issues an understanding that structures themselves were built. Rather than relying entirely on the steam mill, historical actors made industrial capitalism in the process of adopting and applying the steam mill. Historical forces have causes, as market forces have institutional influences. Understanding this process of mutual co-construction, of technology and society, or economics and culture, is the burden of STS that the tobacco story shares.

The contingency and human agency behind technological change make environmental history unappealing. Putting nature so near the center of history has tended to reduce human agency. Examples abound, even among the best practitioners of the discipline. Steven Stoll's excellent *Larding the Lean Earth* (Hill and Wang, 2002) implies that fertilizer from a cow is environmentally superior to fertilizer from a bag. That seems somewhat presentist; such moral calculations rarely framed farmers' economic calculations and technological choices, although sustainability may indeed have been one of their goals, as the rest of Stoll's work demonstrates. More characteristic difficulties in the field appear in Ted Steinberg, *Down to Earth* (Oxford University Press, 2002), which indicated that tobacco's soil-exhausting qualities caused the adoption of slavery in the Chesapeake. Although the field has acknowledged the problems in categorizing some things as natural (see, for example, William Cronon's "The Trouble with Wilderness," in his *Uncommon Ground* [W. W. Norton, 1995]), still many authors ascribe to nature those results that I find more likely caused by the human choices of technique and method. This preference of environmental historians reduces human responsibility for human history. Making nature matter has diminished human accountability. Resorting to nature as the ultimate explanation for historical events explains little.

Since technological choices help build the whole world that seems to derive from natural constraints, numerous other fields and disparate disciplines influenced this work. Historians of science know so much about institution building and recognize the impact of institutions on knowledge and on the human perceptions of the world. Social historians inspire all scholars to put people front and center, even when studying structural change. At the same time, labor historians' long-standing interest in the impact of technological change reminds one to turn those insights inside out and inquire about the racial and sexual divisions of labor as the causes, rather than the effects, of technological change: see Judith A. McGaw, *Most Wonderful Machine* (Princeton University Press, 1987), and Arwen P. Mohun, *Steam Laundries* (Johns Hopkins University Press, 2002). Some cultural histories have come to the same place. Amy Trubek's *Taste of Place* (University of California Press, 2008) examined the relationship between grapes and wine and worked hard to avoid essentializing *terroir* despite recognizing the impact of environment. As my dissertation came to an end, Christophe Lécuyer's *Making Silicon Valley* (MIT Press, 2006) provided inspiration by organizing different chapters with different causative mechanisms—sometimes the market dictated the path of technological development, and sometimes the reverse. Though this book occupies a place in agricultural history and Lecuyer examined informatics, there is a strong connection typical of the history of technology. Subject matter may divide us, but approaches, understandings, and methodologies unite our analyses.

SOUTHERN HISTORIOGRAPHY

This book speaks to a number of very specific debates in Southern history: the origins of the slave system in the British Chesapeake, the compatibility of slavery with antebellum industrialization, the newness (or not) of the New South, similarities and differences between Populists and Progressives, the impact of the New Deal, and the effects of agricultural mechanization, especially after World War II. Yet debates in Southern history often focus on cotton, while shifting our attention to tobacco provides fresh perspectives. As historians tend to do, historical actors associated industrialization with cotton textiles. *DeBow's Review* decried the region's lack of manufacturing in an article that actually followed one about "Southern Negro Life" that dwelt mostly on Richmond's tobacco factories (both, Sept. 1850). Tobacco, in all the complex links of its commodity web, was so well known to these people that they barely noticed it. Historians as well have long associated agriculture with cotton and therefore too closely conflated industrialization with urbanization. It remains mysterious how rigorous economic historians could dismiss the tobacco industry as one that "drew upon agricultural materials for inputs and the farm community for final market," as Fred Bateman and Thomas Weiss did in *A Deplorable Scarcity* (University of North Carolina Press, 1981). Ignoring the facts, they thus denied that Southern manufactures possibly entered a national (let alone an international) market.

For this reason, the history of tobacco technology has the power to address several persistent questions in Southern history: slavery was perfectly compatible with a particular form of industrialization, one whose capital lay more in raw materials than in machines, one that very early relied on branding, marketing, and distribution networks for profit, more than on those efficiencies of production characteristic of Massachusetts textiles. Perhaps putting slaves to work on delicate textile machines would have been a bad idea—resentments could easily throw spanners in such works. We know from sugar production, however, that incentives could ameliorate some of those concerns. The antebellum tobacco industry demonstrates yet another possible model, one too long neglected in the history of American economic growth in the nineteenth century. As for the questions around the adoption of slavery in the Chesapeake, this book argues that colonial inspection laws played a role, though both simple economics and other policies also contributed to the institution, as headrights awarded land to those with the means to supply labor. Only by looking closely at the actual processes of producing tobacco—and avoiding assumptions that agricultural methods represent simple responses to nature or to market demands—did this point emerge. The contextual causes of technological systems reveal so much about their effects.

Likewise, the South after the Civil War underwent and undertook significant changes to the economic and technological systems of agricultural production. Citations in the text can lead students to the scholarship that frames these arguments. Debates about postbellum change versus continuity have reached no definitive conclusion. Structural shifts in the relative value of economic factors (land, labor, and capital) did cause wrenching disruptions. Some planters persisted, while new characters and classes also arose amidst changing conditions, giving force and direction to those transformations. These shifts, however, helped some elites maintain their hold on political and economic power in the region, whether their original fortunes resulted from the war, or emerged after it, or antedated it. While Bright Tobacco (along with cotton sharecropping and tenancy systems) emerged hand in hand with political unrest and upheaval, still the result was a familiar social hierarchy in which race and class informed one another in ways that satisfied and preserved existing power relationships. As postbellum textile industrialization kept the new white working class from exercising political power, so did postbellum tobacco curing methods and credit arrangements keep men like Elias Carr in command of land, labor, and capital—and in charge of government. Although this story is likely told too simply here to satisfy sophisticated Southernists, hopefully they—with other practitioners of the broadest national and regional subfields—will someday incorporate technological change into their more specialized stories.

In Southern history, as well as in other subdisciplines and more general national history too, technological change drops from the sky, or from the discoveries of great men. In other cases, the way things are done looks like the way they have always been done; this has sometimes meant reading postbellum sources to describe antebellum technologies, as Jeffrey Kerr-Ritchie did in the otherwise extraordinary *Freedpeople in the Tobacco South* (University of North Carolina Press, 1999).

To look for the causes of technological change, as well as its effects, is to find a very different picture. When structural changes occur, they play a role in cultural and social transformations; yet structural changes are themselves caused by cultural assumptions and institutional frameworks, as well as changing economic fundamentals and supply-and-demand relationships. In short, the history of technology demonstrates the influence of culture on the large-scale economic structures assumed to control everything else. For that reason, the history of technology matters to many other historical subdisciplines and deserves a closer integration into national narratives as well as economic examinations. Technology has causes as well as effects. Understanding both at the same time is difficult and results in complicated stories, but these are necessary if we wish to grasp and incorporate real history, as it actually happened, instead of the myths of invention more popular in historical analysis.

Index

Page numbers in italics refer to illustrations.